Contents

Nelson Thornes and AQA

Nelson Thornes has worked in partnership with AQA to ensure that this book and the accompanying online resources offer you the best support for your GCSE course.

All AQA endorsed resources undergo a thorough quality assurance process to ensure that their contents closely match the AQA specification. You can be confident that the content of materials branded with AQA's 'Exclusively Endorsed' logo have been written, checked and approved by AQA senior examiners, in order to achieve AQA's exclusive endorsement.

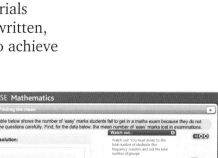

The print and online resources together unlock blended learning; this means that the links between the activities in the book and the activities online blend together to maximise your understanding of a topic and help you achieve your potential.

These online resources are available on *kerboodle!* which can be accessed via the internet at **www.kerboodle.com/live**, anytime, anywhere.

If your school or college subscribes to *kerboodle!* you will be provided with your own personal login details. Once logged in, access your course and locate the required activity.

For more information and help on how to use *kerboodle!* visit **www.kerboodle.com**.

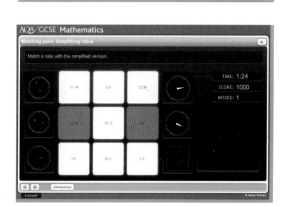

How to use this book

To help you unlock blended learning, we have referenced the activities in this book that have additional online coverage in *kerboodle!* by using this icon:

The icons in this book show you the online resources available from the start of the new specification and will always be relevant.

In addition, to keep the blend up-to-date and engaging, we review customer feedback and may add new content onto *kerboodle!* after publication.

Welcome to GCSE Mathematics

This book has been written by teachers and examiners who not only want you to get the best grade you can in your GCSE exam, but also to enjoy maths. It covers all the material you will need to know for AQA GCSE Mathematics Unit 2 Foundation. This unit does not allow you to use a calculator, so you will not be able to use this most of the time throughout this book. Look out for calculator or non-calculator symbols (shown on the right) which tell you whether to use a calculator or not.

In the exam, you will be tested on the Assessment Objectives (AOs) below. Ask your teacher if you need help to understand what these mean.

AO1 recall and use your knowledge of the prescribed content

AO2 select and apply mathematical methods in a range of contexts

AO3 interpret and analyse problems and generate strategies to solve them.

Each chapter is made up of the following features:

Objectives

The objectives at the start of the chapter give you an idea of what you need to do to get each grade. Remember that the examiners expect you to do well at the lower grade questions on the exam paper in order to get the higher grades. So, even if you are aiming for a Grade C you will still need to do well on the Grade G questions on the exam paper.

On the first page of every chapter, there are also words that you will need to know or understand, called 'Key terms'. The box called 'You should already know' describes the maths that you will have learned before studying this chapter. There is also an interesting fact at the beginning of each chapter which tells you about maths in real life.

 Learn...

The Learn sections give you the key information and examples to show how to do each topic. There are several Learn sections in each chapter.

Practise...

Questions that allow you to practise what you have just learned.

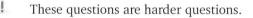

E The bars that run alongside questions in the exercises show you what grade the question is aimed at. This will give you an idea of what grade you're working at. Don't forget, even if you are aiming at a Grade C, you will still need to do well on the Grades G–D questions.

! These questions are harder questions.

⚙ These questions are Functional Maths type questions, which show how maths can be used in real life.

? These questions are problem solving questions, which will require you to think carefully about how best to answer.

🖩 These questions should be attempted **with** a calculator.

�X These questions should be attempted **without** using a calculator.

Assess

End of chapter questions written by examiners. Some chapters feature additional questions taken from real past papers to further your understanding.

Hint

These are tips for you to remember whilst learning the maths or answering questions.

AQA Examiner's tip

These are tips from the people who will mark your exams, giving you advice on things to remember and watch out for.

Bump up your grade

These are tips from the people who will mark your exams, giving you help on how to boost your grade, especially aimed at getting a Grade C.

Consolidation

The consolidation chapter allows you to practise what you have learned in previous chapters. The questions in these chapters can cover any of the topics you have already seen.

1 Types of numbers

Objectives

Examiners would normally expect students who get these grades to be able to:

G

understand place value in large numbers

add and subtract large numbers (up to three digits)

multiply and divide large numbers (up to three digits by two digits)

understand positive and negative integers

find the factors of a number

F

multiply and divide whole numbers by 10, 100, 1000, …

multiply large numbers (two digits by two digits)

add and subtract negative numbers

use inverse operations to check answers

use hierarchy of operations to carry out calculations (BIDMAS)

add and subtract positive and negative numbers

E

multiply and divide positive and negative numbers

recognise prime numbers

C

find the Least Common Multiple (LCM) of two numbers

find the Highest Common Factor (HCF) of two numbers

write a number as a product of prime factors.

Did you know?

Padlocked

Buying on the internet uses **prime numbers**. They keep credit card numbers safe. The product of two very large prime numbers is used as a code. The prime numbers themselves are the key to unlock the code.

The **product** is the result when you multiply numbers together.

Key terms

prime number	positive number
product	negative number
place value	quotient
inverse operation	factor
sum	common factor
difference	highest common factor (HCF)
BIDMAS	multiple
directed number	least common multiple (LCM)
integer	index

You should already know:

✓ how to add, subtract, multiply and divide simple numbers

✓ place value for hundreds, tens and units.

Learn... 1.1 Place value

The value of each digit in a number like 6 349 157 depends on its position:

Six million, three hundred and forty-nine thousand, one hundred and fifty-seven

Thousands						
M	H	T	U	H	T	U
6	3	4	9	1	5	7

6	Millions	6 000 000
3	Hundred Thousands	300 000
4	Ten Thousands	40 000
9	Thousands	9 000
1	Hundreds	100
5	Tens	50
7	Units	7
		6 349 157

Leave spaces (not commas) between groups of three digits.

When writing a number like **sixteen thousand and forty-nine**, take care not to miss out any zeros.

Thousands					
H	T	U	H	T	U
	1	6	0	4	9

AQA *Examiner's tip*

Take care not to miss out zeros like this one – it shows there are no hundreds.

When you **multiply a number by 10**, all the digits **move one place to the left**.

When you **divide a number by 10**, all the digits **move one place to the right**.

For example:

Thousands					
H	T	U	H	T	U
		5	7	9	0
5	7	9	0	0	

5790 × 10 =

The value of each digit is multiplied by 10.

For example,

90 × 10 = 900

An extra zero is needed at the end.

Thousands					
H	T	U	H	T	U
		5	7	9	0
			5	7	9

5790 ÷ 10 =

The value of each digit is divided by 10.

For example,

90 ÷ 10 = 9

The zero disappears.

When you **multiply by 100**, the digits **move two places to the left** (to the right when you divide).

When you **multiply by 1000**, the digits **move three places to the left** (to the right when you divide).

Example: Which is smaller: 50 million or 13 500 000?

Solution:

Millions			Thousands					
H	T	U	H	T	U	H	T	U
	5	0	0	0	0	0	0	0
	1	3	5	0	0	0	0	0

Compare the most significant digit at the front (on the left-hand side).

Writing 50 million in full and comparing it with 13 500 000 shows that:

13 500 000 is smaller.

In words this is thirteen million, five hundred thousand.

It is also equal to thirteen and a half million or 13.5 million.

Practise... 1.1 Place value

G F E D C

G

1 **a** Write down the value of the digit 3 in:

 i 5327 **ii** 9034 **iii** 13 278

 b Write down the value of the digit 7 in:

 i 74 689 **ii** 735 180 **iii** 17 140 094

2 Write these numbers in words:

 a 2746 **f** 90 503

 b 9058 **g** 135 418

 c 9805 **h** 602 700

 d 12 346 **i** 2 045 000

 e 73 060 **j** 35 000 000

3 Write these in figures:

 a eight million

 b eighty thousand

 c eighteen thousand

 d five thousand, four hundred and twenty-nine

 e two thousand and sixty-seven

 f fifteen thousand, nine hundred and seven

 g three hundred and forty-five thousand

 h one hundred and thirty thousand, five hundred

 i six million, seventy-four thousand, nine hundred and five

 j twenty million, four hundred thousand.

> **Hint**
> Take care not to miss out any zeros.

4 Write these numbers in figures in order of size. Start with the smallest:

 7 million one hundred and seventy thousand
 seventeen million 782 000

5 A maths teacher asks her class to write two thousand and forty-seven in figures.

 a Upama writes 247. Write this number in words.

 b Jack writes 200 047. Write this number in words.

 c Write two thousand and forty-seven correctly in figures.

6 **a** **i** Find all the three-figure numbers that can be made using each of the digits 2, 5 and 8 once.

 ii Write the numbers from part **a i** in order of size. Start with the largest.

 b **i** Find all the four-figure numbers that can be made using each of the digits 3, 7, 1 and 6 once.

 ii Write the numbers from part **b i** in order of size. Start with the smallest.

F

7 **a** The number 7320 is multiplied by 10.
 In the new number, what does the digit 3 represent?

 b The number 7320 is divided by 10.
 In the new number, what does the digit 3 represent?

F
E

8 The table gives the heights of some mountains in Europe.

Mountain	Height (m)
Dom	4545
Matterhorn	4479
Mont Blanc	4807
Monte Rosa	4634

a Write the mountains in order of height, starting with the tallest.

b Write the height of the tallest mountain in words.

9 Here are five number cards:

8   4   9   5   1

> **Hint**
>
> **Even** numbers end in 0, 2, 4, 6 or 8.
> **Odd** numbers end in 1, 3, 5, 7 or 9.

a Use all five cards to make the largest possible even number.

b Use all five cards to make the smallest possible odd number.

c Use all five cards to make the number that is nearest to fifty thousand.

d Use all five cards to make the number that is nearest to fifteen thousand.

10 **a** Write these in figures:

   i 425 million

   ii half a million

   iii quarter of a million

   iv 1.5 million

   v 35.6 million

   vi 7 billion

> **Hint**
>
> 1 billion = 1 000 000 000

b Divide each number in part **a** by 10.

c Multiply each number in part **a** by 10.

11 **a** Copy the table. Using the information given, fill in the table using figures.

The population of London is about seven and a half million.

About a quarter of a million people live in Newcastle.

About three quarters of a million people live in Leeds.

The population of Portsmouth is about two hundred thousand.

City	Population
London	
Newcastle	
Leeds	
Portsmouth	

b About one tenth of the UK's population is left-handed. About how many people in each city are left-handed?

> **Hint**
>
> To find one-tenth, divide by 10.

Learn...   1.2 Working with whole numbers

There are sometimes quick ways to add, subtract, multiply or divide whole numbers.

But if you can't see a quick way, you should use one of the standard methods shown below.

They work for any numbers. Start adding, subtracting and multiplying with the units (on the right). Start dividing with the most significant digit (on the left).

You can check each answer by using the **inverse** (the opposite).
Addition and subtraction are inverses.

Multiplication and division are inverses.

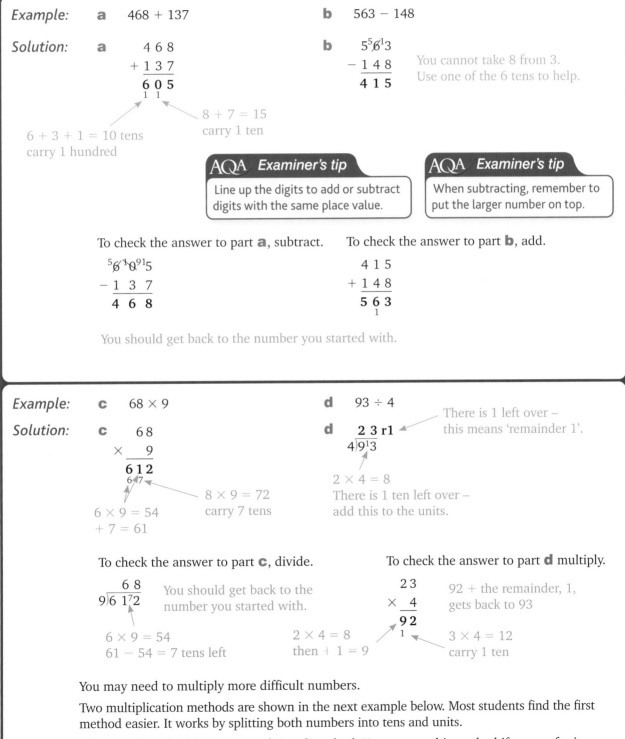

Example: **a**   468 + 137   **b**   563 − 148

Solution: **a**
$$
\begin{array}{r}
4\ 6\ 8 \\
+\ 1\ 3\ 7 \\
\hline
6\ 0\ 5 \\
{\scriptstyle 1\ \ 1}
\end{array}
$$

b
$$
\begin{array}{r}
5\,{}^5\!6\,{}^1 3 \\
-\ 1\ 4\ 8 \\
\hline
4\ 1\ 5
\end{array}
$$
You cannot take 8 from 3. Use one of the 6 tens to help.

8 + 7 = 15
carry 1 ten

6 + 3 + 1 = 10 tens
carry 1 hundred

AQA Examiner's tip
Line up the digits to add or subtract digits with the same place value.

AQA Examiner's tip
When subtracting, remember to put the larger number on top.

To check the answer to part **a**, subtract.
$$
\begin{array}{r}
{}^5\!6\,{}^1\!0\,{}^9 5 \\
-\ 1\ 3\ 7 \\
\hline
4\ 6\ 8
\end{array}
$$

To check the answer to part **b**, add.
$$
\begin{array}{r}
4\ 1\ 5 \\
+\ 1\ 4\ 8 \\
\hline
5\ 6\ 3 \\
{\scriptstyle 1}
\end{array}
$$

You should get back to the number you started with.

Example: **c**   68 × 9   **d**   93 ÷ 4

Solution: **c**
$$
\begin{array}{r}
6\ 8 \\
\times\ \ \ 9 \\
\hline
6\ 1\ 2 \\
{\scriptstyle 6\ 7}
\end{array}
$$

d
$$
\begin{array}{r}
2\ 3\ r1 \\
4\overline{)9\,{}^1 3}
\end{array}
$$
There is 1 left over – this means 'remainder 1'.

2 × 4 = 8
There is 1 ten left over – add this to the units.

8 × 9 = 72
carry 7 tens

6 × 9 = 54
+ 7 = 61

To check the answer to part **c**, divide.
$$
\begin{array}{r}
6\ 8 \\
9\overline{)6\ 1\,{}^7 2}
\end{array}
$$
You should get back to the number you started with.

6 × 9 = 54
61 − 54 = 7 tens left

To check the answer to part **d** multiply.
$$
\begin{array}{r}
2\ 3 \\
\times\ \ \ 4 \\
\hline
9\ 2 \\
{\scriptstyle 1}
\end{array}
$$
92 + the remainder, 1, gets back to 93

2 × 4 = 8
then + 1 = 9

3 × 4 = 12
carry 1 ten

You may need to multiply more difficult numbers.

Two multiplication methods are shown in the next example below. Most students find the first method easier. It works by splitting both numbers into tens and units.

The second method is a more traditional method. You can use this method if you prefer it.

Example: **e**   36 × 24

Solution: **e**

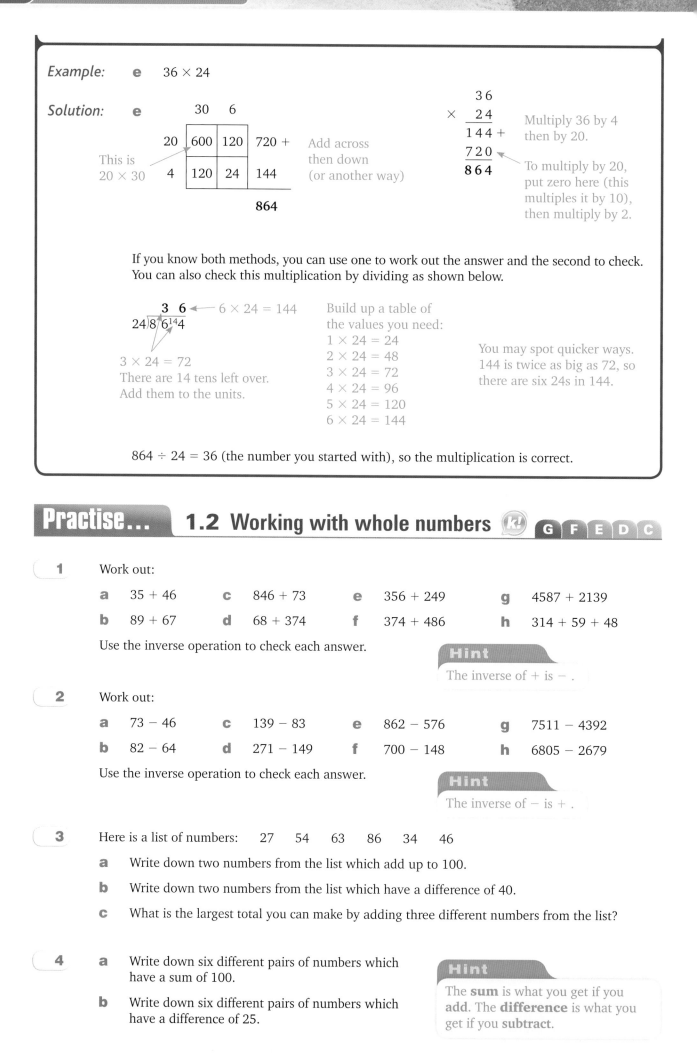

	30	6	
20	600	120	720 +
4	120	24	144

This is 20 × 30

Add across then down (or another way)

864

```
     3 6
  ×  2 4
   1 4 4  +
   7 2 0
   8 6 4
```

Multiply 36 by 4 then by 20.

To multiply by 20, put zero here (this multiples it by 10), then multiply by 2.

If you know both methods, you can use one to work out the answer and the second to check. You can also check this multiplication by dividing as shown below.

```
        3  6  ←── 6 × 24 = 144
   24)8̸ 6¹⁴4
```

3 × 24 = 72
There are 14 tens left over.
Add them to the units.

Build up a table of the values you need:
1 × 24 = 24
2 × 24 = 48
3 × 24 = 72
4 × 24 = 96
5 × 24 = 120
6 × 24 = 144

You may spot quicker ways. 144 is twice as big as 72, so there are six 24s in 144.

864 ÷ 24 = 36 (the number you started with), so the multiplication is correct.

Practise... 1.2 Working with whole numbers 🔑 G F E D C

1   Work out:

a   35 + 46   **c**   846 + 73   **e**   356 + 249   **g**   4587 + 2139

b   89 + 67   **d**   68 + 374   **f**   374 + 486   **h**   314 + 59 + 48

Use the inverse operation to check each answer.

> **Hint**
> The inverse of + is − .

2   Work out:

a   73 − 46   **c**   139 − 83   **e**   862 − 576   **g**   7511 − 4392

b   82 − 64   **d**   271 − 149   **f**   700 − 148   **h**   6805 − 2679

Use the inverse operation to check each answer.

> **Hint**
> The inverse of − is + .

3   Here is a list of numbers: 27   54   63   86   34   46

a   Write down two numbers from the list which add up to 100.

b   Write down two numbers from the list which have a difference of 40.

c   What is the largest total you can make by adding three different numbers from the list?

4   **a**   Write down six different pairs of numbers which have a sum of 100.

    b   Write down six different pairs of numbers which have a difference of 25.

> **Hint**
> The **sum** is what you get if you add. The **difference** is what you get if you **subtract**.

5 Work out:

a	16×5	**f**	136×9	**k**	$135 \div 5$	**p**	$710 \div 9$
b	54×6	**g**	289×4	**l**	$584 \div 2$	**q**	$300 \div 12$
c	35×8	**h**	628×7	**m**	$352 \div 3$	**r**	$775 \div 25$
d	18×7	**i**	$56 \div 4$	**n**	$752 \div 8$	**s**	$810 \div 18$
e	245×3	**j**	$96 \div 6$	**o**	$831 \div 7$	**t**	$578 \div 34$

Use the inverse operation to check your answers.

6 Work out:

a	26×45	**c**	32×83	**e**	52×39	**g**	93×65
b	54×17	**d**	78×41	**f**	86×47	**h**	86×79

Do your answers look reasonable?

7 Given that $8 \times 4 = 32$ and $4 \times 8 = 32$

a write down **two** other multiplications which give the answer 32

b write down the values of   **i** 8×400 **ii** 80×400 **iii** $320 \div 4$

8 Here is a list of numbers: 5   7   8   11   13   36   45

a What is the largest number you can make by multiplying two different numbers on the list?

b What is the smallest number you can make by multiplying three different numbers on the list?

9 **a** Write down six different multiplications that have an answer of 120.

b Write down six different divisions that have an answer of 12.

10 Carol says $806 - 349$ is 543.
Imran says it is 467.
They are both wrong.

a Explain the mistakes they have made.

b What is the correct answer?

Carol
```
  8 0 6
- 3 4 9
-------
  5 4 3
```

Imran
```
 ⁷8¹0¹6
- 3 4 9
-------
  4 6 7
```

11 Sally says 89×7 is 956.
Mike says it is 119.
They are both wrong.

a Find each mistake in their working.

b What is the correct answer?

Sally
```
    8 9
×     7
-------
  9 5 6
  9 3
```

Mike

	80	9	
7	56	63	119

12 Tina thinks $735 \div 7 = 15$

a What mistake has she made?

b Work out the correct answer.

13 You can use these four number cards to make a lot of different four-figure numbers.

   7   4   8   3

a Find the **sum** of the largest possible number and the smallest possible number.

b Find the **difference** between the largest possible number and the smallest possible number.

⚠ 14   Find the missing numbers in these calculations:

a $75 + \boxed{} = 102$

c $\boxed{} - 94 = 428$

e $\boxed{} \div 3 = 29$

b $87 - \boxed{} = 19$

d $\boxed{} \times 4 = 348$

f $96 \div \boxed{} = 8$

⚙ 15   a   The table shows some information about a farm.

   i   What is the total number of animals?

   ii   How many more sheep than cows are there?

   iii   How many more sheep than pigs are there?

Animals on the farm	
Cows	32
Pigs	26
Sheep	83

b   The farmer wants to put a fence around a field.
He plans to leave a gap of 2 metres for a gate.
The farmer has budgeted £5500 for the fence.
The fencing costs £50 for 3 m.
Can the farmer afford the fence he needs?

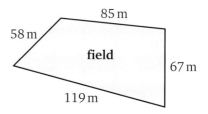

❓ 16   The chart below gives the distances in miles between some cities.
For example, the distance between Derby and York is 88 miles.

Distance chart
Distance in miles

Bristol			
127	**Derby**		
196	70	**Leeds**	
217	88	25	**York**

A salesman lives in Leeds. One day he has to go to Bristol, Derby and York.

a   Find the route with the shortest distance. It must start and end in Leeds.

b   The next day, the salesman has to travel from Leeds to Bristol.
On the way back he has to take a detour.
He checks his mileage for the day.

How many extra miles did his detour add to the journey?

Start   **86 952**

Finish   **87 360**

Learn...   1.3 Order of operations

Some calculations have more than one operation.
The word **BIDMAS** can help you remember the correct order to do them:

B	I	D	M	A	S
Brackets	**Indices**	**Divide**	**Multiply**	**Add**	**Subtract**

These go together.
When they are both in a calculation,
work from left to right.

These go together.
When they are both in a calculation,
work from left to right.

You will meet indices in Chapter 9. For now, be careful to do the other operations in the right order.
The most important thing to remember is to add and subtract last unless they are in a bracket.

For example, in $1 + 2 \times 5$, **multiply** before **adding**. The answer is 11, not 15.

When a calculation is a mixture of adding and subtracting only, you must work from left to right.
This also works when the calculation is a mixture of multiplying and dividing only.

For example, $35 - 8 + 25 = 27 + 25 = 52$ and $28 \div 7 \times 5 = 4 \times 5 = 20$

Example: **a** $9 + 8 \div 4$

$= 9 + 2$ Divide before

$= 11$ adding.

c $3 \times 4 - 8 \div 2$

$= 12 - 4$ Multiply and divide before subtracting.

$= 8$

b $10 \div (2 + 3)$

$= 10 \div 5$ Brackets first.

$= 2$

d $\dfrac{20 - 2 \times 4}{3} = \dfrac{(20 - 2 \times 4)}{3}$ Do the top first – as if it had a bracket.

$= \dfrac{(20 - 8)}{3}$

$= \dfrac{12}{3}$

$= 4$

Practise...   1.3 Order of operations

G F E D C

1   Work out:

a $(7 - 3) \times 6$

b $8 + 4 \div 2$

c $6 \times 5 - 2$

d $9 - (4 + 3)$

e $5 + 3 \times 4$

f $18 \div 2 + 1$

g $7 - (4 + 3)$

h $(2 + 3) \times 5 - 4$

i $15 - 6 \div (2 + 1)$

j $7 \times 2 + 3 \times 5$

k $12 \div (4 - 2)$

l $8 + 6 \div 2 - 1$

2   Say whether each is **true** or **false**. For those that are false, give the correct answer.

a $2 + 4 \times 7 = 42$

b $18 - 4 \div 2 = 16$

c $14 \div (4 + 3) = 2$

d $(9 - 2) \times 4 = 1$

e $9 - (5 - 3) = 1$

f $6 \div (2 + 1) = 4$

g $8 - 4 \div 2 + 1 = 3$

h $6 \times 4 - 2 \times 5 = 14$

3   Work out:

a $\dfrac{8 + 6 \times 2}{5}$

b $\dfrac{4 \times (8 - 5)}{6}$

c $\dfrac{20}{4 + 2 \times 3}$

d $\dfrac{9 \times 4 - 6}{18 - (2 + 1)}$

4   Bob says that $8 - (3 - 2)$ has the same answer as $8 - 3 - 2$.
Is he correct? Explain your answer.

5   Insert brackets to make these correct.

a $2 + 3 \times 4 = 20$

b $15 \div 3 + 2 = 3$

c $12 - 9 - 2 = 5$

d $15 + 7 - 2 \div 5 = 4$

e $15 + 7 - 2 \div 5 = 16$

f $8 - 3 \times 6 - 2 = 20$

6   Copy these. Fill in the missing numbers to make them correct.

a $4 \times 3 + \_\_ = 20$

b $4 \times (3 + \_\_) = 32$

c $(12 - 6) \div \_\_ = 2$

d $12 - 6 \div \_\_ = 9$

e $(12 + 8) \div 4 - \_\_ = 3$

f $12 + 8 \div 4 - \_\_ = 3$

7   Work out:

a $35 \times (52 - 49)$

b $60 + 95 \div 5$

c $(28 + 6) \times 9$

d $\dfrac{9 + 18 \times 7}{15}$

8   Insert brackets to make these correct.

a $70 - 25 + 15 = 30$

b $16 \times 5 - 3 = 32$

c $32 + 16 \div 4 = 12$

d $29 + 37 - 12 \div 6 = 9$

e $72 \div 8 \div 2 = 18$

f $20 - 2 \times 9 + 2 = 198$

F
E

E

9　**a**　Work out $(2 + 7) \times 3$

　　b　Use all of the following to write a single calculation whose answer is as large as possible:
- each of the numbers 4, 5 and 6 (once only)
- each of the operations $+$ and $\times$ (once only)
- one pair of brackets.

10　**a**　Work out $4 + 4 + \frac{4}{4}$

　　b　Can you make each of the numbers 1 to 20 using up to four 4s and mathematical symbols?

11　Use $+ - \times \div$ to make these statements true.

　　a　$2 \_ 3 \_ 4 = 9$

　　b　$2 \_ 3 \_ 4 = 14$

　　c　$2 \_ 3 \_ 4 = 1$

　　d　$2 \_ 3 \_ 4 = 24$

　　e　$2 \_ 3 \_ 4 = 3$

　　f　$2 \_ 3 \_ 4 = 1\frac{1}{2}$

Learn... 1.4 Positive and negative integers

A **directed number** has a positive or negative sign to show whether it is above or below zero.

An **integer** is any positive or negative *whole* number or zero, for example, $-2, -1, 0, +1, +2 \ldots$

Note in **positive numbers** the $+$ signs are often missed out.

You can use a number line to put integers in order.

A number line is like a thermometer standing on its side with ordinary numbers instead of temperatures.

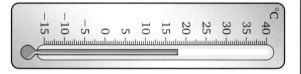

Example: Put these numbers in order of size, smallest first.
9　-13　-37　36　-25　-4　18

Solution: These numbers are shown on the number line.

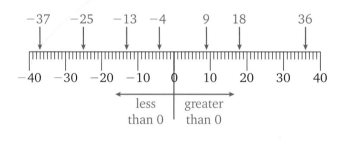

To write them in order, start on the left.

In order the numbers are: -37　-25　-13　-4　9　18　36

Practise...   **1.4 Positive and negative integers** 🇰 G F E D C

G

Use a thermometer or number line to help.

1   Put these temperatures in order. Start with the warmest.

a $-1\,°C$ $2\,°C$ $-2\,°C$ $1\,°C$ $0\,°C$

b $8\,°C$ $-7\,°C$ $6\,°C$ $0\,°C$ $-9\,°C$ $3\,°C$

c $14\,°C$ $-8\,°C$ $5\,°C$ $-10\,°C$ $26\,°C$ $-15\,°C$ $23\,°C$

2   Here are some temperatures.
$9\,°C$ $-4\,°C$ $17\,°C$ $20\,°C$ $-23\,°C$ $-5\,°C$ $-12\,°C$

Write down the temperatures that are

a   higher than $15\,°C$

b   lower than $0\,°C$

c   lower than $-10\,°C$

d   higher than $-5\,°C$.

3   Put these numbers in order, smallest first.

a $+6$ -8 $+5$ $+1$ $+9$ -4 -3

b -16 18 -10 -27 11 0 -19

c 134 -98 47 -103 260 -145 84

4   Here is a list of numbers.
$+2$ -8 $+5$ -1 $+9$ -4 -3

Which numbers are

a   greater than 4

b   less than 0

c   less than -5

d   greater than -3?

5   Judy says that -16 is more than 15. Is she correct?
Explain your answer.

6   Put the correct sign, $<$, $>$ or $=$, between these numbers:

a -5 __ -8

b -4 __ -3

c -5 __ 2

d 7 __ -7

e -9 __ 9

> **Hint**
> $<$ means **less than**. $>$ means **greater than**.

7   **a**   Write down all the integers that lie between -3 and $+6$ (not including -3 and $+6$).

   b   How many integers lie between -87 and -98 (not including -87 and -98)?

Learn...

1.5 Adding and subtracting positive and negative integers

There are two uses for + and − signs: + means **add**
+4 means the positive number 4

− means **subtract**
−4 means the **negative number** −4

You can use a number line to add or subtract positive and negative numbers.

To add numbers – this gives the **sum** of the numbers:
- Find the first number on the number line.
- Go **up (or right)** to add a positive number or **down (or left)** to add a negative number.
- The answer is the number you end on.

AQA *Examiner's tip*

In exams the signs for subtract and negative usually look the same.

To subtract numbers – this gives the **difference** between the numbers:
- Find the first number on the number line.
- Go **down (or left)** to subtract a positive number and **up (or right)** to subtract a negative number. Note that this is the opposite direction to when you are adding.
- The answer is the number you end on.

Examples:

Adding

a $30 + -10 = 20$

b $-5 + 15 = 10$

c $-10 + -20 = -30$

Subtracting

d $10 - 20 = -10$

e $-10 - 20 = -30$

f $-10 - -30 = 20$

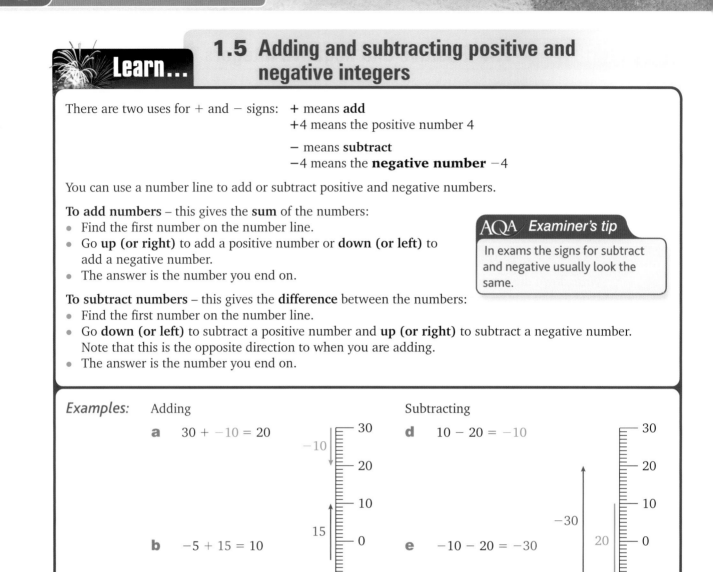

The rules are:

Adding a positive number	+ +	does the same as	+
Adding a negative number	+ −	does the same as	−
Subtracting a positive number	− +	does the same as	−
Subtracting a negative number	− −	does the same as	+

So, for example, $30 + -10 = 30 - 10 = 20$ and $-10 - -30 = -10 + 30 = 20$

Practise...

1.5 Adding and subtracting positive and negative integers

k!

G F E D C

Use a thermometer or a number line to help.

F

1
a The temperature is −3°C. It rises by 4°C. What is the new temperature?

b The temperature is −7°C. It rises by 2°C. What is the new temperature?

c The temperature is 1°C. It falls by 4°C. What is the new temperature?

d The temperature is −2°C. It falls by 3°C. What is the new temperature?

2   Find the value of   **a** $-3 + 6$ **c** $-5 + -3$

                          b $2 + -5$ **d** $-6 + +1$

3   Find the difference between the temperature at midnight and the temperature at midday on each day.

Temperature in a greenhouse		
Day	**Midnight**	**Midday**
Monday	3 °C	12 °C
Tuesday	0 °C	10 °C
Wednesday	−1 °C	7 °C
Thursday	−2 °C	4 °C
Friday	−4 °C	0 °C
Saturday	−6 °C	−1 °C
Sunday	−9 °C	−3 °C

4   Find the value of   **a** $2 - 6$ **c** $-3 - -5$

                          b $-1 - 5$ **d** $-7 - +4$

5   Work out these additions and subtractions. Look at the patterns you get.

    a   **i** $4 + 3$ **iii** $4 + 1$ **v** $4 + -1$ **vii** $4 + -3$

        ii $4 + 2$ **iv** $4 + 0$ **vi** $4 + -2$

    b   **i** $4 - 3$ **iii** $4 - 1$ **v** $4 - -1$ **vii** $4 - -3$

        ii $4 - 2$ **iv** $4 - 0$ **vi** $4 - -2$

6   Find the missing number in each of the following:

    a $-4 + \_\_ = -1$ **e** $1 - \_\_ = -4$

    b $-5 + \_\_ = 0$ **f** $2 - \_\_ = 3$

    c $2 + \_\_ = -3$ **g** $-4 - \_\_ = 0$

    d $-1 + \_\_ = -4$ **h** $-1 - \_\_ = -6$

7   Sam says that $-1 + -5$ has the same answer as $-1 - +5$.
       Is he correct? Explain your answer.

8   Find the next three numbers in each pattern:

    a   6   4   2   0   ....

    b   −11   −8   −5   −2   ....

9   What must be added to:

    a   −2 to make 6   **b**   3 to make −3   **c**   12 to make 7?

10   Work out:

    a $12 - 23$ **e** $-53 - 24$

    b $-15 - 19$ **f** $29 - -46$

    c $28 - -35$ **g** $72 - 98$

    d $-21 - -34$ **h** $-45 - -37$

11   Some cities and their temperatures at midnight are shown on the map.

   a   Which city is the warmest?

   b   Which city is the coldest?

   c   Find the difference in temperature between:

     i   London and Vienna

     ii   London and Rome

     iii   Brussels and Oslo

     iv   Madrid and Moscow.

Midnight temperatures

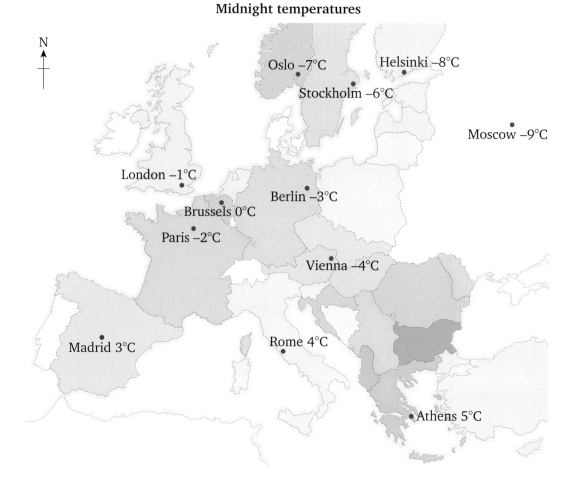

   d   In London, the temperature rises 3 °C by 6 am.
      What is the temperature at 6 am?

   e   In Stockholm the temperature falls 2 °C by 6 am.
      What is the temperature at 6 am?

12

ENGLISH HILLS AND LAKES			
Heights of hills		**Depths of lakes**	
Catstycam	+889 m	Coniston	−56 m
Great Dodd	+857 m	Grasmere	−25 m
Helvellyn	+951 m	Ullswater	−60 m
Scafell Pike	+978 m	Wastwater	−79 m

   a   How much higher is Catstycam than Great Dodd?

   b   How much deeper is Coniston than Grasmere?

   c   Find the difference in height between:

     i   the top of Helvellyn and the bottom of Ullswater

     ii   the bottom of Wastwater and the top of Scafell Pike.

? 13 Each number in a magic square is different.

The sum of each row, each column and each diagonal is the same.

0		4
	1	−3
−2		

a Copy this magic square and fill in the missing numbers.

b If you add or subtract the same number to each box you get another magic square. Explain why.

c Make another magic square.

d Check that it works, then give a copy to a friend to solve.

? 14 Three whole numbers have a sum of 10.

a One of the numbers is 14. Give an example of what the other two numbers might be.

b One of the numbers is 16 and the other two numbers are equal. What are they?

Learn... 1.6 Multiplying and dividing positive and negative integers

You can write multiplications as additions. For example, 5×2 is the same as adding 5 lots of 2.

a $+5 \times +2 = +2 + +2 + +2 + +2 + 2 = +10$ and $+5 \times +2 = +5 + +5 = +10$

b $+5 \times -2 = -2 + -2 + -2 + -2 - 2 = -10$ and $-2 \times -5 = -5 + -5 = -10$

      5 lots of -2 2 lots of -5

These show that multiplying one positive by one negative number gives a negative answer.

a and **b** show that changing the sign of one of the numbers changes the sign of the answer.

Starting with $+5 \times -2 = -10$ and changing $+5$ to -5,

the result will be $-5 \times -2 = +10$

This shows that multiplying two negative numbers together gives a positive answer.

The results for muliplication are also true for division. So when **multiplying or dividing**, remember:

When signs are the **same**	$+ \times +$ or $- \times -$ $+ \div +$ or $- \div -$	the answer is positive $+$
When signs are **different**	$+ \times -$ or $- \times +$ $+ \div -$ or $- \div +$	the answer is negative $-$

Example:

a $+2 \times +5 = +10$ This is the same as 2×5 Decide on the sign of the answer, then multiply the numbers.

b $+4 \times -3 = -12$

c $-7 \times +5 = -35$ Signs different gives negative

d $-3 \times -2 = +6$ Signs same gives positive

e $+12 \div +3 = +4$ This is the same as $12 \div 3$

f $+15 \div -5 = -3$ Signs different gives negative

g $-20 \div +4 = -5$

h $-18 \div -2 = +9$ Signs same gives positive

> **AQA Examiner's tip**
>
> Take care! Multiplying and dividing follow these rules, but adding and subtracting don't.
> For example, $-3 + 5 = 2$ NOT -2.
> Remember to use a number line for adding and subtracting.

The answer when you multiply numbers is called the **product**.

The answer when you divide numbers is called the **quotient**.

Practise...

1.6 Multiplying and dividing positive and negative integers

G F E D C

E

1 Work out:

a $+3 \times +5$
b $+2 \times -6$
c $-8 \times +5$
d -7×-2

e $+5 \times +7$
f $+4 \times -8$
g $-6 \times +7$
h -9×-8

2 Work out:

a $+12 \div +3$
b $+16 \div -2$
c $-30 \div +5$
d $-32 \div -4$

e $+42 \div +6$
f $+63 \div -7$
g $-64 \div +8$
h $-36 \div -9$

3 Which of these are correct? If the answer is wrong, write down the correct answer.

a $+3 \times +4 = +12$
b $+2 \times -8 = +16$
c $-4 \times +5 = -20$
d $-6 \times -2 = -12$
e $4 \times 7 = -28$
f $4 \times -6 = -24$
g $-9 \times 7 = 63$
h $-7 \times -8 = 56$

i $+9 \div +3 = -27$
j $+8 \div -2 = -4$
k $-15 \div +5 = +3$
l $-20 \div -4 = -5$
m $36 \div 6 = -6$
n $-56 \div 7 = -8$
o $-48 \div 8 = -6$
p $-54 \div -9 = 6$

D

4 Find the missing number in each of the following.

a $5 \times \underline{\ \ } = -25$
b $-2 \times \underline{\ \ } = 14$
c $-4 \times \underline{\ \ } = 0$
d $-3 \times \underline{\ \ } = -24$

e $-16 \div \underline{\ \ } = -2$
f $27 \div \underline{\ \ } = -3$
g $-7 \div \underline{\ \ } = 7$
h $24 \div \underline{\ \ } = -6$

5 Ian says that $+7 \times -5$ has the same answer as $-7 \times +5$.
Is he correct? Explain your answer.

6 a Find all the pairs of integers that have a product of -16.

b Find all the pairs of integers that have a product of 20.

7 Find the values of these:

a $\dfrac{-5 \times +4}{+2}$
b $\dfrac{+6 \times -4}{-2}$
c $\dfrac{-2 \times -9}{+3}$
d $\dfrac{-8 \times -3}{-4}$

8 Work out:

a -25×4
b 6×-47
c -32×-80
d 26×-15

e $-94 \div 2$
f $-72 \div -3$
g $275 \div -5$
h $-432 \div -9$

9 A quiz has 12 questions.
Contestants get two points for a correct answer.
One point is taken off for each wrong answer.

a Find the missing points (*) in this table of results.

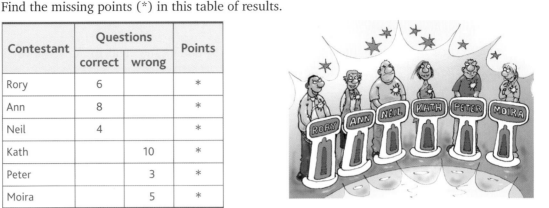

Contestant	Questions		Points
	correct	wrong	
Rory	6		*
Ann	8		*
Neil	4		*
Kath		10	*
Peter		3	*
Moira		5	*

b Find **i** the highest possible number of points
     ii the lowest possible number of points.

c How many correct answers give

   i 3 points   **ii** −9 points   **iii** 18 points   **iv** −3 points?

10 a Find two integers whose sum is −8 and whose product is 12.

   b Find two integers whose sum is −7 and whose product is –18.

   c Find two **negative** integers whose difference is 5 and whose product is 24.

Make up integer descriptions of your own. Ask a friend to find the integers.

Learn... 1.7 Factors and multiples

A **factor** is a positive whole number that divides exactly into another number.
For example, the factors of 16 are   1, 2, 4, 8, 16

Factors usually occur in pairs:
$1 \times 16 = 16, 2 \times 8 = 16, 4 \times 4 = 16$

A factor is sometimes called a divisor.

To find all the factors of a number, look for factor pairs.

For example,   20 = 1 × 20   so 1 and 20 are factors of 20
               20 = 2 × 10   so 2 and 10 are factors of 20
               20 = 4 × 5   so 4 and 5 are factors of 20.

The factors of 20 are 1, 2, 4, 5, 10, 20.

The **common factors** of two or more numbers are the factors that
they have in common.

The **highest common factor (HCF)** of two or more numbers is
the highest factor that they have in common.

AQA *Examiner's tip*

Be systematic so you don't lose
any factors.

AQA *Examiner's tip*

Remember, 1 is a factor of all
numbers.

Example: Find the highest common factor (HCF) of 20 and 24.

Solution: A factor is something that goes into the number.

The factors of 20 are **1**, **2**, **4**, 5, 10, 20.
The factors of 24 are **1**, **2**, 3, **4**, 6, 8, 12, 24.

The common factors are the numbers that are in both lists.

The common factors are **1, 2, 4.**
The highest common factor is **4.**

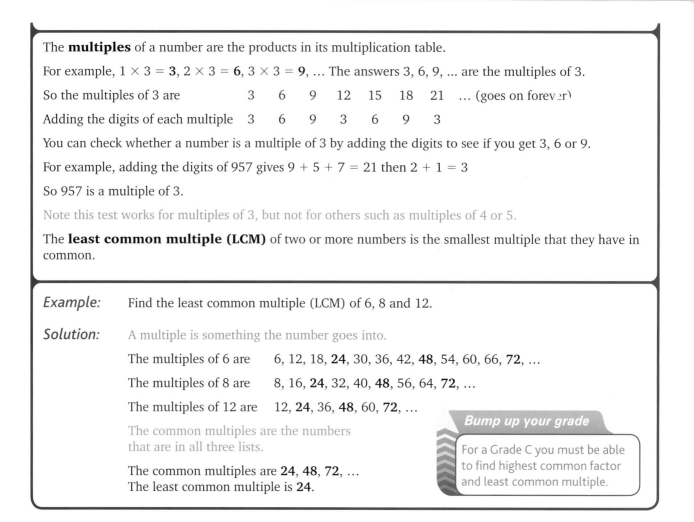

The **multiples** of a number are the products in its multiplication table.

For example, $1 \times 3 = 3$, $2 \times 3 = 6$, $3 \times 3 = 9$, ... The answers 3, 6, 9, ... are the multiples of 3.

So the multiples of 3 are   3   6   9   12   15   18   21   ... (goes on forever)

Adding the digits of each multiple   3   6   9   3   6   9   3

You can check whether a number is a multiple of 3 by adding the digits to see if you get 3, 6 or 9.

For example, adding the digits of 957 gives $9 + 5 + 7 = 21$ then $2 + 1 = 3$

So 957 is a multiple of 3.

Note this test works for multiples of 3, but not for others such as multiples of 4 or 5.

The **least common multiple (LCM)** of two or more numbers is the smallest multiple that they have in common.

Example: Find the least common multiple (LCM) of 6, 8 and 12.

Solution: A multiple is something the number goes into.

The multiples of 6 are   6, 12, 18, **24**, 30, 36, 42, **48**, 54, 60, 66, **72**, ...

The multiples of 8 are   8, 16, **24**, 32, 40, **48**, 56, 64, **72**, ...

The multiples of 12 are   12, **24**, 36, **48**, 60, **72**, ...

The common multiples are the numbers that are in all three lists.

The common multiples are **24, 48, 72**, ...
The least common multiple is **24**.

Bump up your grade

For a Grade C you must be able to find highest common factor and least common multiple.

Practise...   1.7 Factors and multiples 🎧   G F E D C

G

1   Here is a list of numbers: 2   3   4   5   6   7

Which of these numbers are factors of:

a   8   **c**   20   **e**   84

b   12   **d**   28   **f**   420?

2   Find all the factors of:

a   4   **c**   10   **e**   25

b   9   **d**   18   **f**   60

3   Write down the first five multiples of:

a   4   **b**   6   **c**   7   **d**   8

4   **a**   **i**   Write down the first six multiples of 10.

  ii   How can you tell whether a number is a multiple of 10?

  b   **i**   Write down the first six multiples of 5.

  ii   How can you tell whether a number is a multiple of 5?

  c   Here is a list of numbers: 90   105   210   306   495   570

  Which of these are multiples of: **i**   10   **ii**   5   **iii**   3?

D

5 **a** Write down the first 12 multiples of 9. Then find their digit sums.

b Here is a list of numbers: 153   207   378   452   574   3789
Which of these do you think are multiples of 9?

c Check your answers to part **b** by dividing by 9.

d How can you tell whether a number is a multiple of 9?

6 Find the common factors of:

a 6 and 9 **c** 14 and 35 **e** 15 and 35

b 8 and 28 **d** 24 and 36 **f** 12 and 30.

C

7 Find the common factors of 42 and 70 and write down the highest common factor.

8 Find the factors then the highest common factor of the following pairs of numbers:

a 6 and 15 **c** 24 and 32 **e** 50 and 75

b 12 and 18 **d** 48 and 60 **f** 42 and 70.

9 The highest common factor of two numbers is 7. Give three possible pairs of numbers.

10 Find the least common multiple of the following sets of numbers:

a 4 and 6 **d** 12 and 20

b 7 and 5 **e** 2, 3 and 5

c 6 and 8 **f** 3, 4 and 5.

11 Tracy says that the least common multiple of 12 and 30 is 6.
Is she correct? Explain your answer.

12 Find the highest common factor of:

a 6, 18 and 24 **b** 36, 45 and 54 **c** 14, 56 and 84.

13 Find the least common multiple of the following sets of numbers:

a 8, 10 and 12 **b** 6, 8 and 32 **c** 15, 20 and 25.

14 A planet has moons called Titania and Oberon.
Titania goes round the planet every 9 days.
Oberon goes round the planet every 13 days.
On one evening the moons are in line with the planet.
How many days will it be before they are in line again?

15 **a** A shop is open every day. The farm delivers milk to it every 2 days and butter every 3 days. Today they had a delivery of milk and butter. How many days will it be before the deliveries **next** arrive on the **same** day?

b A different shop is open every day. The farm delivers milk to it every 2 days, butter every 3 days and eggs every 7 days. Today they had a delivery of milk, butter and eggs. How many days will it be before all the deliveries **next** arrive on the **same** day?

Learn... 1.8 Prime numbers and prime factors

A prime number is a positive whole number that has **exactly two factors**.

The first 7 prime numbers are:

2	3	5	7	11	13	17
Factors	Factors	Factors	Factors	Factors	Factors	Factors
1 & 2	1 & 3	1 & 5	1 & 7	1 & 11	1 & 13	1 & 17

1 is not a prime number because it has only one factor.

2 is the only even prime number. All other even numbers have factors of 1, themselves and 2.

All the other missing odd numbers have three or more factors.
For example, the factors of 15 are 1, 3, 5 and 15.

Index form

Prime numbers are the 'building blocks' of mathematics.
All other numbers can be written as products of prime numbers.

This is sometimes called prime factor decomposition.

For example, $12 = 2 \times 2 \times 3 = 2^2 \times 3$

product   **index**

and $81 = 9 \times 9 = 3 \times 3 \times 3 \times 3 = 3^4$

This is called index form.   index

It is more difficult to find the prime factors of larger numbers.

You can use the **tree method** to do this. See the example below.

Example: Write 280 as a product of its prime factors in index form.

Solution: Two 'trees' are shown below. The first starts by splitting 280 into 28×10.

The numbers are then split again and again until you get to prime numbers.

The second tree starts by splitting 280 into 2×140.

This shows that whichever tree you use, you end with three 2s, a 5 and a 7.

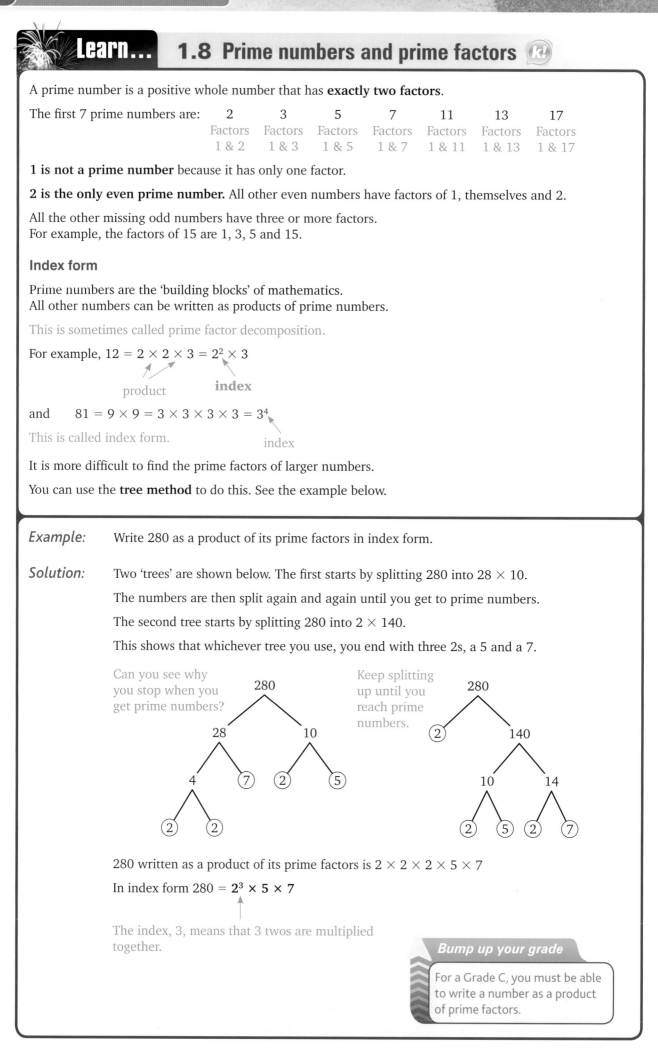

Can you see why you stop when you get prime numbers?

Keep splitting up until you reach prime numbers.

280 written as a product of its prime factors is $2 \times 2 \times 2 \times 5 \times 7$

In index form $280 = \mathbf{2^3 \times 5 \times 7}$

The index, 3, means that 3 twos are multiplied together.

Bump up your grade

For a Grade C, you must be able to write a number as a product of prime factors.

Practise...

1.8 Prime numbers and prime factors

k!

G F E D C

D

C

1 Write down all the prime numbers between 20 and 30.

2 Which of these numbers are **not** prime numbers? 31 33 35 37 39 41
Explain your answers.

3 Write each number as a product of prime factors.
  a 14   **b** 30   **c** 33   **d** 42   **e** 65   **f** 91

4 Write each number as a product of prime factors.
Write your answers using index notation.
  a 24   **b** 36   **c** 45   **d** 64   **e** 84   **f** 96

5 Ruth says that if you write 40 as a product of prime factors, the answer is $1 \times 2^3 \times 5$.
Is she correct? Explain your answer.

6 Write each number as a product of prime factors. Use index notation.
  a 100   **b** 132   **c** 144   **d** 153   **e** 216   **f** 520

7 Which of these numbers **cannot** be prime? 895 356 3457 5739
Explain your answers.

8 **a** Write each number as a product of prime factors, using index notation.
      i 27   **ii** 45
  b Use your answers to part **a** to find:
      i the HCF of 27 and 45   **ii** the LCM of 27 and 45.

9 **a** Write each number as a product of prime factors, using index notation.
      i 42   **ii** 60   **iii** 72
  b Use your answers to part **a** to find:
      i the HCF of 42, 60 and 72   **ii** the LCM of 42, 60 and 72.

10 The product of two prime numbers is sometimes used as a security device.
To 'break the code' you need to find two prime numbers that give a particular product.

Find two prime numbers that multiply to give:
  a 111   **b** 221   **c** 319   **d** 437   **e** 767

Why are even numbers not very useful in this situation?

11 Find the mystery number in each part.
  a It is a prime number. It is a factor of 35. It is not a factor of 25.
  b It is less than 50. It is a multiple of 3. It is also a multiple of 5.
The sum of its digits is a prime number.
  c It is a prime number less than 100.
It is one more than a multiple of 8 and its digits add up to 10.

Make up number descriptions of your own. Ask a friend to find the numbers.

Assess (k!)

G

1 The price of a new car is £18 490.

a Write the number 18 490 in words.

b In the number 18 490, write down the value of

   i the digit 9   **ii** the digit 8.

c The number 18 490 is divided by 10.
   In the new number, what does the digit 8 represent?

2 Work out:

a 354 + 289   **c** 86 × 7   **e** 245 ÷ 5

b 600 − 138   **d** 46 × 29   **f** 900 ÷ 8

> **Hint**
> You could use inverse operations or your calculator to check.

3 Write down all the factors of 40.

G E

4 Here is a list of numbers: 3   5   6   8   16   23   27

From this list, write down:

a two numbers that add up to 31   **d** a factor of 9

b two numbers that have a difference of 7   **e** three numbers with a product of 240

c a multiple of 9   **f** three prime numbers.

F E

5 The table below gives the temperatures of some cities at dawn.

City	Birmingham	Bristol	Leeds	Manchester	Newcastle
Temperature	−1 °C	2 °C	−4 °C	−3 °C	−5 °C

a Which city has   **i** the lowest temperature   **ii** the highest temperature?

b Which cities have temperatures below −1 °C?

6 Copy and complete the following table.

Temperature	Change	New temperature
3 °C	+4 °C	
2 °C	−5 °C	
−4 °C	+9 °C	
−1 °C	−5 °C	
7 °C		11 °C
12 °C		8 °C
−3 °C		7 °C
−1 °C		−4 °C
	+6 °C	9 °C
	−5 °C	4 °C
	+13 °C	10 °C
	−2 °C	−7 °C

7 The first five terms of a sequence are 160, −80, 40, −20.

Find the next four terms.

8 a Find the highest common factor of 45 and 75.

b Find the least common multiple of 45 and 75.

9 Write 392 as a product of its prime factors in index form.

10 James races two model cars around a track.
The first car takes 42 seconds to complete each circuit.
The second car takes 1 minute to complete each circuit.
The cars start together from the starting line.
How long will it be before they are together on the
starting line again?

AQA Examination-style questions

1 Here is a list of numbers:

3 4 6 8 9 12 18

    a Write down **four** different numbers from the list that add up to 30. *(1 mark)*

    b Write down **one** number in the list that is a multiple of 6. *(1 mark)*

    c Write down **all** the numbers in the list that are factors of 18. *(2 marks)*

    d There are two square numbers in the list. Work out the difference between them. *(2 marks)*

AQA 2008

2 Find a multiple of 4 and a multiple of 5 that add to make a multiple of 6. *(2 marks)*

AQA 2008

3 a Write 36 as the product of prime factors.
Give your answer in index form. *(3 marks)*

    b What is the Least Common Multiple (LCM) of 12 and 36? *(1 mark)*

AQA 2008

2   Sequences

Examiners would normally expect students who get these grades to be able to:

G

continue a sequence of diagrams or numbers

write the terms of a simple sequence

F

find a term in a sequence with positive numbers

write the term-to-term rule in a sequence with positive numbers

E

find a term in a sequence with negative or fractional numbers

write the term-to-term rule in a sequence with negative or fractional numbers

D

write the terms of a sequence or a series of diagrams given the nth term

C

write the nth term of a linear sequence or a series of diagrams.

Key terms

sequence
ascending
descending
term-to-term
linear sequence
nth term

Did you know?

Sequences in nature

Have you ever wondered why four-leaf clovers are so rare? It's because four isn't a number in the Fibonacci sequence.

The Fibonacci sequence 0, 1, 1, 2, 3, 5, 8, 13, ... is well known in nature and can be applied to seashell shapes, branching plants, flower petals, pine cones and pineapples.

If you count the number of petals on a daisy, you are most likely to find 13, 21, 34, 55 or 89 petals. These are all numbers in the Fibonacci sequence.

You should already know:

✓ how to identify odd and even numbers.

Learn... 2.1 The rules of a sequence

A **sequence** is a set of numbers or patterns with a given rule.

2, 4, 6, 8, 10, ... is the sequence of even numbers

The rule is **add two**.

Sequences can be

- **ascending** (going up) e.g. 2, 5, 8, 11, ...
- **descending** (going down) e.g. 5, 3, 1, −1, −3, ...

Sequences can also be patterns.

The rule is **add one square**.

Example: Write down the next two terms in this sequence.

5, 10, 15, 20, ... The dots tell you that the sequence continues.

Solution: The rule to find the next number in the sequence is **+5**.

1st   2nd   3rd   4th
term   term   term   term

5,   10,   15,   20,

+5   +5   +5

The rule (called the **term-to-term** rule) can be used to find the next two terms.

The fifth term is $20 + 5 = 25$

The sixth term is $25 + 5 = 30$

5,   10,   15,   20,   25,   30,

+5   +5   +5   +5   +5

Example: Write down the next two terms in this sequence.

5, 10, 20, 40, ...

Solution: The rule to find the next number in the sequence is $\times 2$.

1st   2nd   3rd   4th
term   term   term   term

5,   10,   20,   40,

×2   ×2   ×2

The fifth term is $40 \times 2 = 80$

The sixth term is $80 \times 2 = 160$

5,   10,   20,   40,   80,   160,

×2   ×2   ×2   ×2   ×2

Practise... 2.1 The rules of a sequence (k!)   G F E D C

G

1 Draw the next two diagrams in the following sequences.

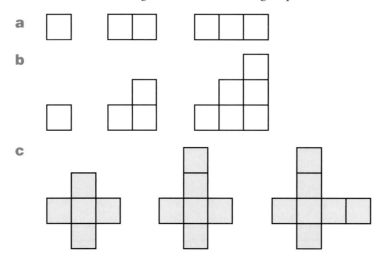

a

b

c

F

2 Write down the next two terms of the following sequences.

a   3, 7, 11, 15, ...

b   4, 8, 12, 16, ...

c   2, 10, 18, 26, ...

d   0, 5, 10, 15, ...

e   1, 2, 4, 8, 16, ...

f   100, 1000, 10 000, 100 000, ...

E

3 Copy and complete the following table.

Pattern (n)	Diagram	Number of matchsticks (m)
1		3
2		5
3		7
4		
5		

a   What do you notice about the pattern of matchsticks above?

b   How many matchsticks will there be in the fifth pattern?

c   How many matchsticks will there be in the tenth pattern?

d   There are 41 matchsticks in the 20th pattern.
How many matchsticks are there in the 21st pattern?
Give a reason for your answer.

4 Write down the term-to-term rule for the following sequences.

a 3, 7, 11, 15, ...

b 0, 5, 10, 15, ...

c 1, 2, 4, 8, 16, ...

d 3, 4.5, 6, 7.5, ...

e 20, 16, 12, 8, ...

f 100, 1000, 10 000, ...

g 2, 3, 4.5, 6.75, ...

h 54, 18, 6, 2, ...

5 The term-to-term rule is +4.
Write down five different sequences that fit this rule.

6 Here is a sequence of numbers.

3   5   9   17

The rule for continuing this sequence is: multiply by 2 and subtract 1.

a What are the next two numbers in this sequence?

The same rule is used for a sequence that starts with the number −3.

b What are the first four numbers in this sequence?

7 Here is a sequence of coordinates:

(2, 5), (3, 6), (4, 7), ...

What are the next two coordinates in this sequence?
What do you notice if you plot these points on a graph?

8 Jacob is exploring number patterns.

He writes down the following products in a table.

1 × 1	1
11 × 11	121
111 × 111	12 321
1111 × 1111	1 234 321
11 111 × 11 111	
111 111 × 111 111	

a Copy and complete the next two rows of the table.

b Jacob says he can use the table to work out 1 111 111 111 × 1 111 111 111
Is he correct? Give a reason for your answer.

9 Here are the first four terms of a sequence.

45, 31, 14, 15

Here is the rule for the sequence.
To get the next number, multiply the digits of the previous number and add 11 to the result.
Work out the 100th number of the sequence.

10 Investigate the following sequences. What are the rules and next terms?

a 1, 4, 9, 16, ...

b 1, 8, 27, 64, ...

c 1, 3, 6, 10, 15, ...

d 1, 1, 2, 3, 5, 8, ...

e 2, 3, 5, 7, 11, ...

Learn... 2.2 The *n*th term of a sequence

A **linear sequence** is one where the differences between the terms are all the same.

The sequence 5, 10, 15, 20, … is a linear sequence because the differences are all the same.

5,　　10,　　15,　　20,
　　$+5$　$+5$　$+5$

The sequence 5, 10, 20, 40, … is NOT a linear sequence because the differences are not the same.

5,　　10,　　20,　　40,
　　$+5$　$+10$　$+20$

To find the **nth term** of a linear sequence, you can use the formula:

*n*th term = difference $\times$ *n* + (first term $-$ difference)

$= dn + (a - d)$

For 7, 10, 13, 16 *d* is the difference = $+3$

　　　　　　　　a is the first term = 7

*n*th term = difference $\times$ *n* + (first term $-$ difference)

$= 3 \times n + (7 - 3)$

$= 3n + 4$

Example:　The *n*th term of a sequence is $2n + 3$

Find the first four terms of the sequence.

Solution:　1st term $= 2 \times 1 + 3 = 5$

2nd term $= 2 \times 2 + 3 = 7$

3rd term $= 2 \times 3 + 3 = 9$　　Similarly

4th term $= 2 \times 4 + 3 = 11$　　100th term $= 2 \times 100 + 3 = 203$

The first four terms are 5, 7, 9, 11.

The sequence is called a linear sequence because the differences between the terms are all the same.

5,　　7,　　9,　　11, …
　$+2$　$+2$　$+2$

In this example, the differences are all $+2$.

The term-to-term rule is $+2$.

Example:　The first four terms of a sequence are 7, 10, 13, 16.

Find the *n*th term.

Solution:　7,　　10,　　13,　　16,
　　　　　$+3$　$+3$　$+3$

The term-to-term rule is $+3$.

This tells you that the rule is of the form $3n + …$

The sequence goes up in 3s, just like the 3 times table, so the rule begins $3 \times n$ (3*n*, for short).

1st term $= 3 \times 1 + … = 7$

2nd term $= 3 \times 2 + … = 10$

3rd term $= 3 \times 3 + … = 13$

4th term $= 3 \times 1 + … = 16$

From the above you can see that the *n*th term is $3n + 4$

$= 3n + (7 - 3)$　This method only works for linear sequences.

$= 3n + 4$

Practise...   2.2 The *n*th term of a sequence   G F E D C

D

1 Write down the first five terms of the sequence whose *n*th term is:

   a $n + 5$ **c** $5n + 1$ **e** $n^2 + 2$ **g** $10 - 2n$

   b $3n$ **d** $2n - 7$ **f** $\frac{1}{2}n + 2$

2 Write down the 100th and the 101st terms of the sequence whose *n*th term is:

   a $n + 3$ **b** $3n - 10$ **c** $100 - 2n$ **d** $n^2 - 1$

3 Aisha writes down the sequence 2, 6, 10, 14, ...
She says that the *n*th term is $n + 4$

Is she correct?
Give a reason for your answer.

4 The *n*th term of a sequence is $3n - 1$

   a Colin says that 31 is a number in this sequence.
      Is Colin correct? Give a reason for your answer.

   b Diane says the 20th term is double the 10th term.
      Is Diana correct? Give a reason for your answer.

5 Write down the *n*th term for the following linear sequences.

   a 3, 7, 11, 15, ...   **e** −1, 1, 3, 5, ...

   b 0, 6, 12, 18, ...   **f** −5, −1, 3, 7, ...

   c 9, 15, 21, 27, ...   **g** 5, 6.5, 8, 9.5, ...

   d 8, 14, 20, 26, ...   **h** 23, 21, 19, 17, ...

6 Write down the 10th and the 100th terms of each of the sequences in Question 5.

C

7

Pattern (*n*)	Diagram	Number of matchsticks (*m*)
1		3
2		5
3	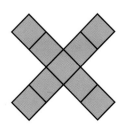	7

   a Write down the formula for the number of matchsticks (*m*) in the *n*th pattern.

   b There are 200 matchsticks. What pattern number can be made?

8 Write down the formula for the number of squares in the *n*th pattern.

> **AQA** *Examiner's tip*
> Always check your *n*th term to see that it works for the sequence.

C

⚠ 9   Stuart says that the number of cubes in the 100th pattern is 300.

How can you tell Stuart is wrong?

Give a reason for your answer.

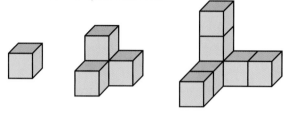

⚠ 10   Write the *n*th term for these non-linear sequences.

a   1, 4, 9, 16, …   d   1, 8, 27, 64, 125, …

b   2, 5, 10, 17, …   e   0, 7, 26, 63, 124, …

c   2, 8, 18, 32, …   f   10, 100, 1000, …

> **Hint**
>
> Use your answer to part **a** to help you with parts **b** and **c**. Use your answer to part **d** to help you with parts **e** and **f**.

⚠ 11   Write the *n*th term for the following sequences.

a $1 \times 2, 2 \times 3, 3 \times 4, \ldots$

b $\frac{2}{3}, \frac{3}{4}, \frac{4}{5}, \frac{5}{6}, \ldots$

c $1 \times 2 \times 5, 2 \times 3 \times 6, 3 \times 4 \times 7, 4 \times 5 \times 8, \ldots$

d   0.1, 0.2, 0.3, 0.4, …

⚙ 12   Jackie builds fencing from pieces of wood as shown below.

Diagram 1   **Diagram 2**   **Diagram 3**
4 pieces of wood   7 pieces of wood   10 pieces of wood

a   How many pieces of wood will there be in Diagram *n*?

b   Use your answer to part **a** to work out the number of pieces of wood needed for Diagram 10.

⚙ 13   The table shows the stopping distances for cars travelling at different speeds.

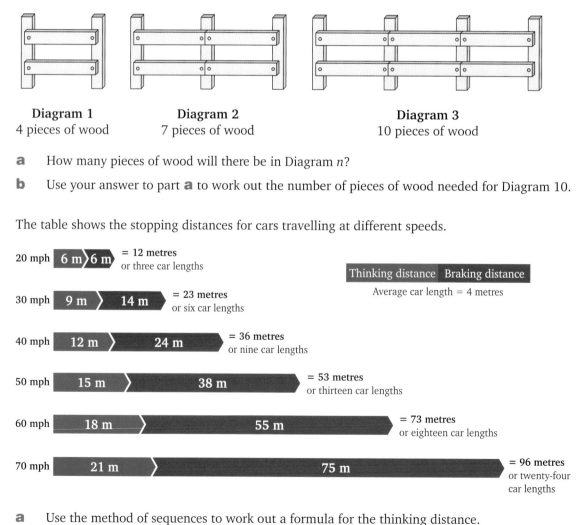

a   Use the method of sequences to work out a formula for the thinking distance.

b   Use the formula to work out the thinking distance for **i** 80 mph and **ii** 35 mph.

2 Assess 𝑘!

1   **a**   Draw the next two diagrams in the following sequences.

   i

   ii

   iii

   b   Write down the next two terms in the following sequences.

   i   2, 7, 12, 17, …

   ii   1.2, 1.4, 1.6, 1.8, …

   iii   2, 6, 18, 54, …

   iv   5, 50, 500, 5000, …

2   **a**   Write down the 5th, 8th and 10th terms in the following sequences.

   i   2, 5, 8, 11, …

   ii   1, 6, 11, 16, …

   iii   3, 6, 12, 24, …

   b   Write down the term-to-term rule in the following sequences.

   i   1, 5, 9, 13, …

   ii   2, 10, 50, 250, …

   iii   1, 2, 5, 14, 41, …

3   **a**   Write down the 5th, 8th and 10th terms in the following sequences.

   i   20, 17, 14, 11, …

   ii   64, 32, 16, …

   iii   1, −2, 4, −8, …

   b   Write down the term-to-term rule for the following sequences.

   i   8, 5, 2, −1, …

   ii   −1, −4, −7, −10, …

   iii   2, −2, 2, −2, 2, …

4   The nth term of a sequence is $4n + 3$

   Asha says that the 10th term is double the 5th term.

   Is Asha correct?

   Give reasons for your answer.

5   Information about some squares is shown.

   a   Copy and complete the table.

   b   Use the table to write down the nth term in these sequences.

   i   2, 8, 18, 32, …

   ii   −2, 1, 6, 13, …

Side of square (cm)	1	2	3	4	n
Area of square (cm²)	1	4	9	16	

G
F
F
E
D
D

C

6   Find the nth term in the following sequences.

     a   6, 8, 10, 12, …

     b   3, 13, 23, 3, …

     c   8, 6, 4, 2, …

7   The nth term of a sequence is $4n - 5$

     The nth term of a different sequence is $8 + 2n$

     Jo says that there are no numbers that are in both sequences.

     Show that Jo is correct.

8   Write down a linear sequence where the 4th term is twice the 2nd term.

     Jo says that that this is always true if the first term is equal to the difference.

     Is Jo correct?

     Give a reason for your answer.

AQA Examination-style questions

1   **a**   The first term of a sequence is -2.
            The rule for continuing the sequence is:

> Add 7
> then multiply by 4

        What is the second term of the sequence?   *(1 mark)*

     b   This rule is used to continue a different sequence.

> Multiply by 2
> then add 5

        The **third** term of this sequence is 11.
        Work out the **first** term.   *(4 marks)*

AQA 2005

3 Fractions

$$\cfrac{1}{1+\cfrac{1}{1+\cfrac{1}{1+1}}}$$

Did you know?

Continuing fractions

The word fraction comes from the Latin word *frangere* meaning 'to break into pieces'.

Here's an amazing fraction!

$$\cfrac{1}{1+\cfrac{1}{1+\cfrac{1}{1+1}}}$$

It is part of a series of fractions that starts:

$$1,\ \cfrac{1}{1+1},\ \cfrac{1}{1+\cfrac{1}{1+1}},\ \cfrac{1}{1+\cfrac{1}{1+\cfrac{1}{1+1}}},\ \cfrac{1}{1+\cfrac{1}{1+\cfrac{1}{1+\cfrac{1}{1+1}}}}$$

This is a sequence of what are called continuing fractions. Can you see how the sequence could continue?

The fractions in the sequence simplify to $1, \frac{1}{2}, \frac{2}{3}, \frac{3}{5}, \frac{5}{8}, \ldots$ Can you continue this sequence?

Do you recognise these numbers?

You may not be able to do it yet, but by the end of the chapter perhaps you will!

Key terms

equivalent fractions
numerator
denominator
mixed number
integer
improper fraction
reciprocal

You should already know:

✔ how to add, subtract, multiply and divide simple numbers

✔ the meaning of 'sum', 'difference' and product'

✔ how to use simple fractions such as halves and quarters.

Learn... 3.1 Simple fractions

You can think of a simple fraction in different ways. For example, $\frac{3}{4}$ is

- a position on a number line halfway between a half and one:

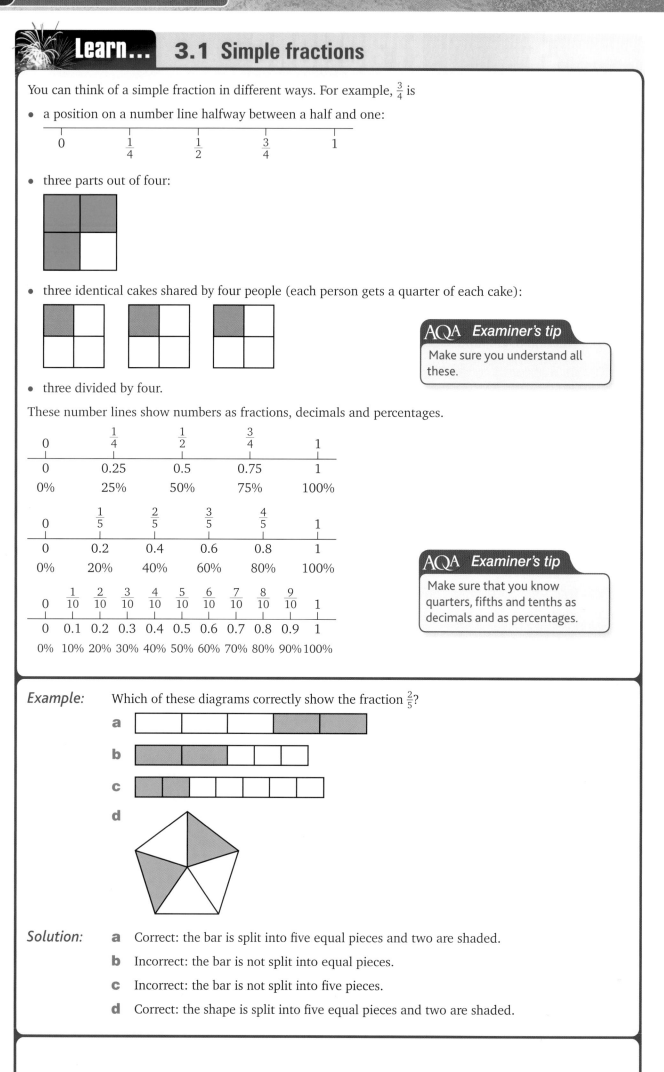

- three parts out of four:

- three identical cakes shared by four people (each person gets a quarter of each cake):

> **AQA** *Examiner's tip*
>
> Make sure you understand all these.

- three divided by four.

These number lines show numbers as fractions, decimals and percentages.

> **AQA** *Examiner's tip*
>
> Make sure that you know quarters, fifths and tenths as decimals and as percentages.

Example: Which of these diagrams correctly show the fraction $\frac{2}{5}$?

a

b

c

d

Solution:

a Correct: the bar is split into five equal pieces and two are shaded.

b Incorrect: the bar is not split into equal pieces.

c Incorrect: the bar is not split into five pieces.

d Correct: the shape is split into five equal pieces and two are shaded.

Example: Write these mixed numbers as decimals: $1\frac{1}{2}$, $2\frac{2}{5}$, $3\frac{1}{4}$, $4\frac{7}{10}$

Solution: $1\frac{1}{2} = 1.5$

The integer part stays the same. The fraction part is replaced by the matching decimal. (Use the number lines if you need to.)

$2\frac{2}{5} = 2.4$

$3\frac{1}{4} = 3.25$

$4\frac{7}{10} = 4.7$

Example: Arrange these fractions in order, starting with the lowest: $\frac{2}{5}$, $\frac{3}{4}$, $\frac{1}{2}$, $\frac{3}{10}$

Solution: One way of doing this is to write each fraction as a decimal.

They are 0.4, 0.75, 0.5, 0.3, which in order is 0.3, 0.4, 0.5, 0.75.

So the fractions in order are: $\frac{3}{10}$, $\frac{2}{5}$, $\frac{1}{2}$, $\frac{3}{4}$

Practise...   3.1 Simple fractions

G F E D C

1   Copy the grids.
Shade one-quarter in four different ways.

G

2   At the beginning of this chapter are different ways of seeing the fraction $\frac{3}{4}$
Do the same for the fraction $\frac{7}{10}$

3   Draw diagrams to show that:

$\frac{1}{4}$ is the same as $\frac{2}{8}$

$\frac{4}{5}$ is the same as $\frac{8}{10}$

$\frac{2}{3}$ is the same as $\frac{6}{9}$

4   Write these percentages as fractions.

a	30%	**c**	75%	**e**	90%
b	80%	**d**	20%		

Hint

Use the number lines if you do not yet know the fractions.

5   Write these fractions as decimals.

a	$\frac{1}{2}$	**c**	$\frac{4}{5}$	**e**	$1\frac{9}{10}$	**g**	$\frac{12}{10}$
b	$\frac{3}{4}$	**d**	$\frac{3}{10}$	**f**	$2\frac{3}{5}$	**h**	$\frac{11}{5}$

Try to do it without using the number lines.

F

6   Write these fractions as percentages and then arrange them in order of size, starting with the smallest.

a	$\frac{3}{5}$	**b**	$\frac{4}{5}$	**c**	$\frac{1}{2}$	**d**	$\frac{3}{4}$	**e**	$\frac{3}{10}$

E

7 **a** Which fraction is bigger, one-fifth or one-quarter?
How do you know?

b Which fraction is bigger, four-fifths or three-quarters?
How do you know?

c Arrange the fractions one-fifth, one-quarter, four-fifths, three-quarters in order, starting with the smallest.

8 **a** Which fraction is bigger, $\frac{1}{8}$ or $\frac{1}{9}$? How do you know?

b Which fraction is bigger, $\frac{7}{8}$ or $\frac{8}{9}$? How do you know?

c Arrange these fractions in order of size, starting with the smallest.
$\frac{7}{8}$, $\frac{1}{2}$, $\frac{8}{9}$, $\frac{1}{8}$, $\frac{1}{9}$

9 Write down a fraction that is between:

a a half and a quarter

b two-thirds and one

c zero and one-ninth.

10 Dave says '$\frac{4}{7}$ is smaller than $\frac{5}{7}$ because 4 is smaller than 5.'

Is Dave correct? Explain why.

James says, '$\frac{3}{4}$ is smaller than $\frac{3}{5}$ because 4 is smaller than 5.'

Is James correct? Explain why.

11 Kiki has drunk one-quarter of a bottle of juice.
Lulu has drunk two-tenths of a bottle the same size.
Who has drunk more? Show how you worked it out.

Learn... 3.2 Equivalent fractions

In the previous section you saw some fractions that are the same in value, such as $\frac{1}{2}$ and $\frac{4}{8}$.

The diagram shows that $\frac{1}{2}$ and $\frac{4}{8}$ are the same.

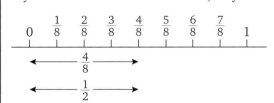

Four parts out of eight (four-eighths) are shaded.

Also, the left part is shaded; the right is not. So, one part out of two, one-half, is shaded.

Fractions such as $\frac{1}{2}$ and $\frac{4}{8}$ with the same value are called **equivalent fractions**.

If you mark them on a number line, they are at the same point.

$$0 \quad \frac{1}{8} \quad \frac{2}{8} \quad \frac{3}{8} \quad \frac{4}{8} \quad \frac{5}{8} \quad \frac{6}{8} \quad \frac{7}{8} \quad 1$$

$$\longleftarrow \frac{4}{8} \longrightarrow$$

$$\longleftarrow \frac{1}{2} \longrightarrow$$

Other fractions equivalent to $\frac{1}{2}$ are $\frac{2}{4}$, $\frac{3}{6}$, $\frac{4}{8}$, $\frac{5}{10}$, ...

Note that in each case the **numerator** is half the **denominator**.

Any fraction with numerator half the denominator is equivalent to one-half.

Equivalent fractions are useful for adding and subtracting fractions (see Learn 3.3) and for putting them in order of size.

This diagram shows a number line from 0 to 1 split up into thirds and fifteenths.

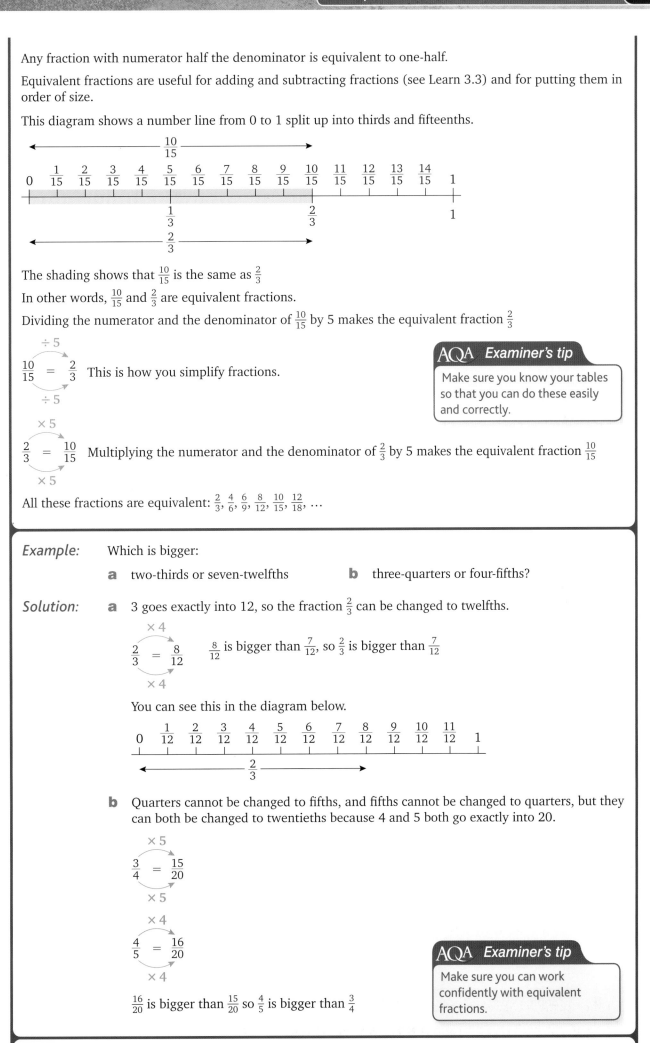

The shading shows that $\frac{10}{15}$ is the same as $\frac{2}{3}$

In other words, $\frac{10}{15}$ and $\frac{2}{3}$ are equivalent fractions.

Dividing the numerator and the denominator of $\frac{10}{15}$ by 5 makes the equivalent fraction $\frac{2}{3}$

$$\frac{10}{15} = \frac{2}{3}$$ This is how you simplify fractions.

$\div 5$

AQA *Examiner's tip*

Make sure you know your tables so that you can do these easily and correctly.

$\times 5$

$$\frac{2}{3} = \frac{10}{15}$$ Multiplying the numerator and the denominator of $\frac{2}{3}$ by 5 makes the equivalent fraction $\frac{10}{15}$

$\times 5$

All these fractions are equivalent: $\frac{2}{3}, \frac{4}{6}, \frac{6}{9}, \frac{8}{12}, \frac{10}{15}, \frac{12}{18}, \ldots$

Example: Which is bigger:

    a   two-thirds or seven-twelfths   **b**   three-quarters or four-fifths?

Solution: **a**   3 goes exactly into 12, so the fraction $\frac{2}{3}$ can be changed to twelfths.

    $\times 4$

    $\frac{2}{3} = \frac{8}{12}$ $\frac{8}{12}$ is bigger than $\frac{7}{12}$, so $\frac{2}{3}$ is bigger than $\frac{7}{12}$

    $\times 4$

You can see this in the diagram below.

    b   Quarters cannot be changed to fifths, and fifths cannot be changed to quarters, but they can both be changed to twentieths because 4 and 5 both go exactly into 20.

    $\times 5$

    $\frac{3}{4} = \frac{15}{20}$

    $\times 5$

    $\times 4$

    $\frac{4}{5} = \frac{16}{20}$

    $\times 4$

AQA *Examiner's tip*

Make sure you can work confidently with equivalent fractions.

$\frac{16}{20}$ is bigger than $\frac{15}{20}$ so $\frac{4}{5}$ is bigger than $\frac{3}{4}$

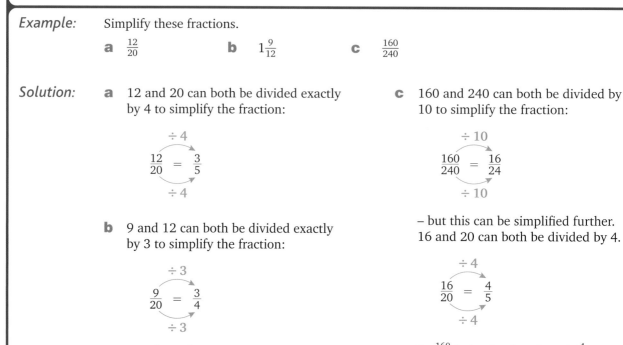

Example: Simplify these fractions.

a $\frac{12}{20}$ b $1\frac{9}{12}$ c $\frac{160}{240}$

Solution:

a 12 and 20 can both be divided exactly by 4 to simplify the fraction:

$$\frac{12}{20} = \frac{3}{5}$$
$\div 4$

b 9 and 12 can both be divided exactly by 3 to simplify the fraction:

$$\frac{9}{20} = \frac{3}{4}$$
$\div 3$

So $1\frac{9}{12} = 1\frac{3}{4}$

c 160 and 240 can both be divided by 10 to simplify the fraction:

$$\frac{160}{240} = \frac{16}{24}$$
$\div 10$

– but this can be simplified further. 16 and 20 can both be divided by 4.

$$\frac{16}{20} = \frac{4}{5}$$
$\div 4$

So $\frac{160}{240}$ in its simplest form is $\frac{4}{5}$.

Practise... 3.2 Equivalent fractions 🔘

G F E D C

G

1 Write each fraction in its simplest form.

a $\frac{6}{8}$ c $\frac{8}{10}$ e $\frac{24}{36}$ g $\frac{48}{56}$ i $\frac{35}{80}$ k $\frac{64}{100}$

b $\frac{10}{20}$ d $\frac{30}{100}$ f $\frac{32}{40}$ h $\frac{76}{84}$ j $\frac{150}{250}$ l $\frac{108}{144}$

2 Complete each pair of equivalent fractions.

a $\frac{3}{11} = \frac{}{22}$ c $\frac{3}{} = \frac{15}{25}$ e $\frac{}{15} = \frac{9}{45}$ g $\frac{7}{10} = \frac{}{150}$

b $\frac{8}{9} = \frac{24}{}$ d $\frac{}{20} = \frac{85}{100}$ f $\frac{12}{} = \frac{48}{80}$

3 Use this fraction wall to write down three equivalent fraction facts.

$\frac{1}{9}$	$\frac{1}{9}$	$\frac{1}{9}$	$\frac{1}{9}$	$\frac{1}{9}$	$\frac{1}{9}$	$\frac{1}{9}$	$\frac{1}{9}$	$\frac{1}{9}$

$\frac{1}{6}$	$\frac{1}{6}$	$\frac{1}{6}$	$\frac{1}{6}$	$\frac{1}{6}$	$\frac{1}{6}$

$\frac{1}{3}$	$\frac{1}{3}$	$\frac{1}{3}$

$\frac{1}{2}$	$\frac{1}{2}$

4 Write down three fractions equivalent to each of:

a $\frac{3}{4}$ b $\frac{2}{3}$ c $\frac{5}{6}$

F

5 Which of these are equivalent to $\frac{4}{5}$?

a $\frac{12}{15}$ b $\frac{40}{50}$ c $\frac{9}{10}$ d $\frac{45}{55}$ e $\frac{120}{150}$ f $\frac{90}{100}$

E

6 Sophie says, '$\frac{0.4}{0.5} = \frac{4}{5}$'

Is she correct? Explain your answer.

7 What fraction is the odd one out in each list?

a $\frac{2}{3}$ $\frac{3}{4}$ $\frac{4}{6}$ $\frac{6}{9}$

b $\frac{3}{5}$ $\frac{6}{10}$ $\frac{16}{20}$ $\frac{12}{20}$

E

8 Amy says these fractions are all equivalent.

$\frac{4}{7}$ $\frac{5}{8}$ $\frac{6}{9}$ $\frac{7}{10}$ $\frac{8}{11}$

Is Amy correct? Explain your answer.

9 Arrange these fractions in order.

a $\frac{5}{6}, \frac{3}{18}, \frac{1}{3}, \frac{4}{9}$

Hint
Change the fractions to eighteenths to compare them.

b $\frac{5}{12}, \frac{3}{8}, \frac{3}{4}, \frac{1}{6}$

10 Complete these:

a $\frac{1}{2} = \frac{a}{\rule{1em}{0.4pt}} = \frac{2b}{\rule{1em}{0.4pt}} = \frac{}{2x}$

b $\frac{2}{5} = \frac{2a}{\rule{1em}{0.4pt}} = \frac{}{5y} = \frac{4x}{\rule{1em}{0.4pt}} = \frac{}{10t}$

11 How many slices are there in a quarter of each pizza?

a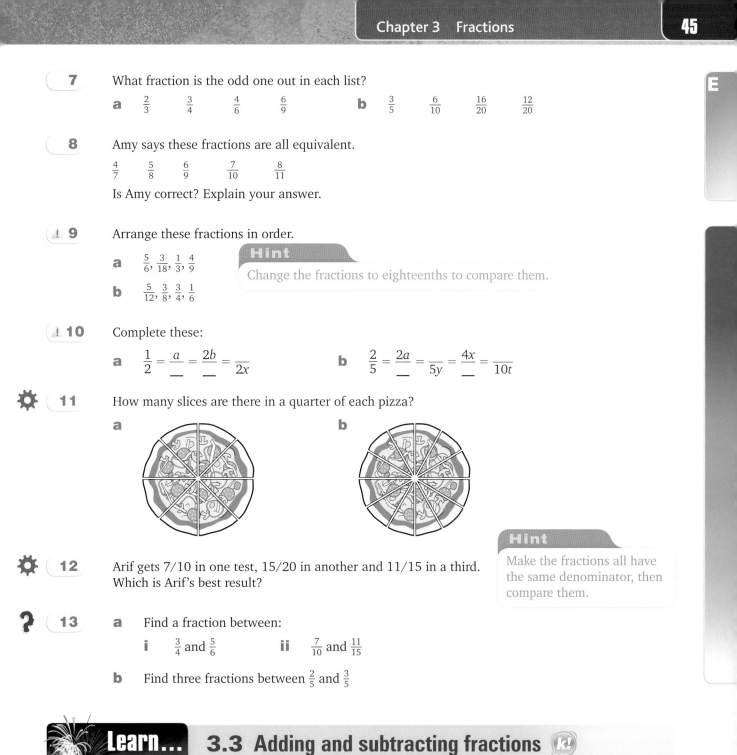

b

12 Arif gets 7/10 in one test, 15/20 in another and 11/15 in a third. Which is Arif's best result?

Hint
Make the fractions all have the same denominator, then compare them.

13 a Find a fraction between:

i $\frac{3}{4}$ and $\frac{5}{6}$

ii $\frac{7}{10}$ and $\frac{11}{15}$

b Find three fractions between $\frac{2}{5}$ and $\frac{3}{5}$

Learn... **3.3 Adding and subtracting fractions**

To add fractions with different denominators you first have to change them so that they have the same denominator.

To find the sum of three-quarters and two-thirds you have to change the fractions to twelfths, because 12 is the smallest number that is a multiple of both 3 and 4.

$\frac{3}{4}$ is $\frac{9}{12}$

$\frac{2}{3}$ is $\frac{8}{12}$

Hint
24, or any other multiple of 12, would also have worked here, but 12 is better because it is smaller, and so easier to use. You can always find a number that is a multiple of two numbers by multiplying them together. (Here, 3 × 4 = 12)

So $\frac{3}{4} + \frac{2}{3} = \frac{9}{12} + \frac{8}{12} = \frac{17}{12} = 1\frac{5}{12}$

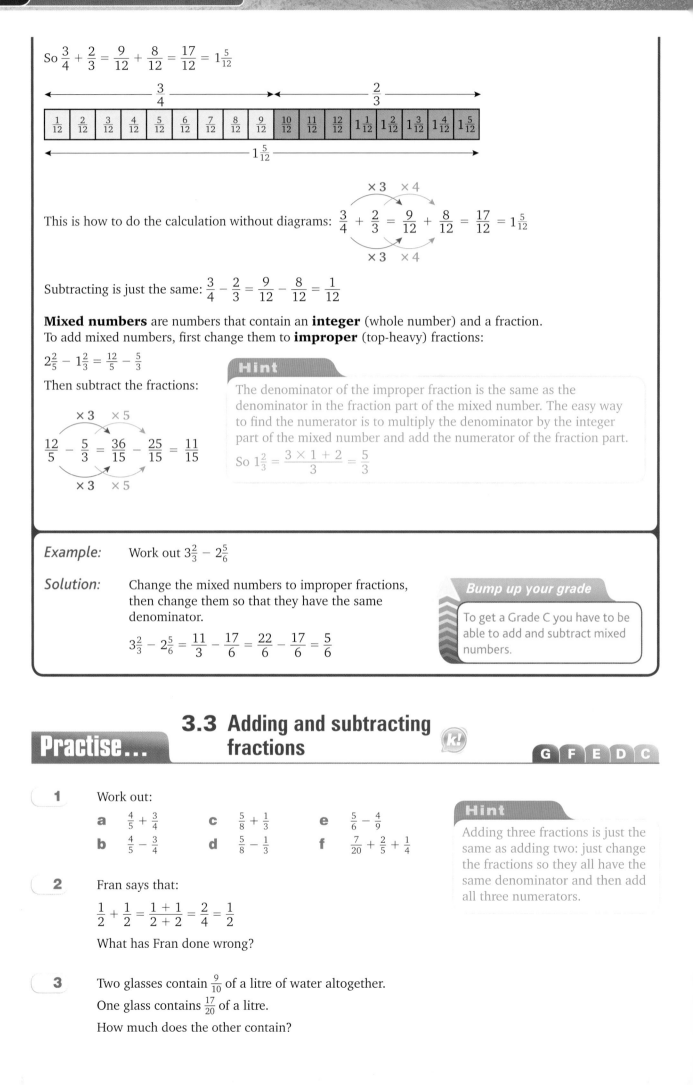

This is how to do the calculation without diagrams: $\frac{3}{4} + \frac{2}{3} = \frac{9}{12} + \frac{8}{12} = \frac{17}{12} = 1\frac{5}{12}$

Subtracting is just the same: $\frac{3}{4} - \frac{2}{3} = \frac{9}{12} - \frac{8}{12} = \frac{1}{12}$

Mixed numbers are numbers that contain an **integer** (whole number) and a fraction. To add mixed numbers, first change them to **improper** (top-heavy) fractions:

$2\frac{2}{5} - 1\frac{2}{3} = \frac{12}{5} - \frac{5}{3}$

Then subtract the fractions:

$\frac{12}{5} - \frac{5}{3} = \frac{36}{15} - \frac{25}{15} = \frac{11}{15}$

Hint

The denominator of the improper fraction is the same as the denominator in the fraction part of the mixed number. The easy way to find the numerator is to multiply the denominator by the integer part of the mixed number and add the numerator of the fraction part.

So $1\frac{2}{3} = \frac{3 \times 1 + 2}{3} = \frac{5}{3}$

Example: Work out $3\frac{2}{3} - 2\frac{5}{6}$

Solution: Change the mixed numbers to improper fractions, then change them so that they have the same denominator.

$3\frac{2}{3} - 2\frac{5}{6} = \frac{11}{3} - \frac{17}{6} = \frac{22}{6} - \frac{17}{6} = \frac{5}{6}$

Bump up your grade

To get a Grade C you have to be able to add and subtract mixed numbers.

3.3 Adding and subtracting fractions

Practise... $\quad$ k! $\quad$ G F E D C

D

1 Work out:

a $\frac{4}{5} + \frac{3}{4}$ $\qquad$ c $\frac{5}{8} + \frac{1}{3}$ $\qquad$ e $\frac{5}{6} - \frac{4}{9}$

b $\frac{4}{5} - \frac{3}{4}$ $\qquad$ d $\frac{5}{8} - \frac{1}{3}$ $\qquad$ f $\frac{7}{20} + \frac{2}{5} + \frac{1}{4}$

Hint

Adding three fractions is just the same as adding two: just change the fractions so they all have the same denominator and then add all three numerators.

2 Fran says that:

$\frac{1}{2} + \frac{1}{2} = \frac{1+1}{2+2} = \frac{2}{4} = \frac{1}{2}$

What has Fran done wrong?

3 Two glasses contain $\frac{9}{10}$ of a litre of water altogether.

One glass contains $\frac{17}{20}$ of a litre.

How much does the other contain?

4   Four of these calculations give the same answer and one gives a different
answer. Which is the odd one out? Show how you worked it out.

$\frac{1}{2} + \frac{1}{3} + \frac{1}{6}$

$\frac{1}{2} + \frac{1}{2}$

$\frac{2}{3} + \frac{1}{4} + \frac{1}{12}$

$\frac{3}{4} + \frac{1}{8} + \frac{1}{16}$

$\frac{3}{5} + \frac{1}{3} + \frac{1}{15}$

D

5   Work out:

a $\frac{3}{4} + \frac{3}{4}$ **c** $\frac{3}{4} + \frac{3}{4} + \frac{3}{4} + \frac{3}{4}$

b $\frac{3}{4} + \frac{3}{4} + \frac{3}{4}$ **d** $\frac{3}{4} + \frac{3}{4} + \frac{3}{4} + \frac{3}{4} + \frac{3}{4}$

6   Work out:

a $3\frac{3}{4} + 1\frac{4}{5}$ **b** $3\frac{3}{4} - 1\frac{4}{5}$

C

7   Anne's recipe needs $\frac{2}{3}$ of a cup of sugar. She has $\frac{3}{4}$ of a cup.
How much will she have left?

June's recipes need $1\frac{1}{2}$ cups of sugar and $1\frac{2}{3}$ cups of sugar.
How much sugar does June need altogether?

8   (In America, fabric is sold in yards and fractions of a yard rather than in metres
and tenths of a metre as in the UK.)

A pair of trousers needs $1\frac{1}{2}$ yards of fabric and a jacket needs $2\frac{3}{8}$ yards.
How much fabric is needed in total?

9   Amy mixes a fruit drink. It is made up of $\frac{1}{3}$ of a litre of orange juice, $\frac{2}{5}$ of a litre
of apple juice and $\frac{7}{10}$ of a litre of blackcurrant juice. How much fruit drink is
there in total? Amy adds some water to make it up to 3 litres.
How much water has she added?

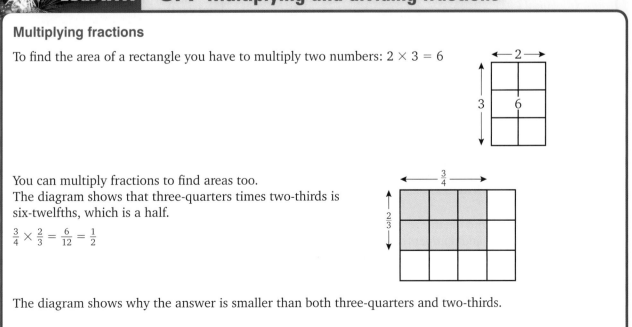

Learn...   3.4 Multiplying and dividing fractions

Multiplying fractions

To find the area of a rectangle you have to multiply two numbers: $2 \times 3 = 6$

You can multiply fractions to find areas too.
The diagram shows that three-quarters times two-thirds is
six-twelfths, which is a half.

$\frac{3}{4} \times \frac{2}{3} = \frac{6}{12} = \frac{1}{2}$

The diagram shows why the answer is smaller than both three-quarters and two-thirds.

The 6 in the sixth-twelfths comes from the shaded area – it is three units across and two units down, so contains six squares.

The 12 in the answer comes from the whole area – it is four units across and three units down, so contains 12 squares.

Six out of 12 squares are shaded, so the answer is $\frac{6}{12}$, which simplifies to $\frac{1}{2}$

multiply numerators $\div 6$

This is how to do it without the diagram: $\quad \frac{3}{4} \times \frac{2}{3} = \frac{3 \times 2}{4 \times 3} = \frac{6}{12} = \frac{1}{2}$

multiply denominators $\div 6$

You can simplify before working out: $\quad \frac{3}{4} \times \frac{2}{3} = \frac{{}^1\!3 \times 2^1}{{}_2\!4 \times 3_1} = \frac{1 \times 1}{2 \times 1} = \frac{1}{2}$

AQA *Examiner's tip*

Be careful not to mix up the method for adding fractions and the method for multiplying.

What about calculations such as $1\frac{3}{4} \times 2\frac{2}{3}$?

You can do this by changing the mixed numbers to improper fractions:

$1\frac{3}{4} \times 2\frac{2}{3} = \frac{7}{4} \times \frac{8}{3}$

Then simplify if possible and multiply to get the answer.

$1\frac{3}{4} \times 2\frac{2}{3} = \frac{7}{{}_1\!4} \times \frac{8^2}{3} = \frac{7}{1} \times \frac{2}{3} = \frac{14}{3} = 4\frac{2}{3}$

If the calculation involves integers, just write the integer as a fraction with a denominator of 1.

$5 \times 4\frac{1}{3} = \frac{5}{1} \times \frac{13}{4} = \frac{65}{4} = 16\frac{1}{4}$

Division of fractions

To divide by a fraction, you multiply by its **reciprocal.**

Dividing by $\frac{3}{4}$ is the same as multiplying by $\frac{4}{3}$:

$\frac{2}{3} \div \frac{3}{4} = \frac{2}{3} \times \frac{4}{3} = \frac{8}{9}$

Hint

You find the reciprocal of a fraction by turning it upside down.
So the reciprocal of $\frac{3}{4}$ is $\frac{4}{3}$.

For mixed numbers, change them to improper fractions first:

$1\frac{3}{4} \div 2\frac{2}{3} = \frac{7}{4} \div \frac{8}{3}$

Then turn the second fraction to its reciprocal and multiply:

$\frac{7}{4} \times \frac{3}{8} = \frac{21}{32}$

AQA *Examiner's tip*

Make sure you understand all these fraction ideas so that you do not get them mixed up in the exam.

Example: Work out:

a $3\frac{1}{2} \times 2\frac{5}{6}$

b $4 \div 3\frac{2}{3}$

Solution: **a**   Change the mixed numbers to improper fractions, then multiply numerators and denominators.

$3\frac{1}{2} \times 2\frac{5}{6} = \frac{7}{2} \times \frac{17}{6} = \frac{119}{12} = 9\frac{11}{12}$

b   Write the integer as a fraction with a denominator of 1, then change the mixed number to an improper fraction. Change dividing to multiplying by the reciprocal, then multiply numerators and denominators.

$4 \div 3\frac{2}{3} = \frac{4}{1} \div \frac{11}{3} = \frac{4}{1} \times \frac{3}{11} = \frac{12}{11} = 1\frac{1}{11}$

Practise... 3.4 Multiplying and dividing fractions k! G F E D C

D

1 Work out:

a   i $\frac{2}{5} \div \frac{1}{3}$ ii $\frac{1}{3} \div \frac{2}{5}$

b   What do you notice about the answers?

c   Try another pair of fractions and compare answers.

d   Does the same thing happen with $5 \div 3$ and $3 \div 5$?

2 The reciprocal of a fraction is $3\frac{7}{8}$. What is the fraction?

3 These calculations can all be done very easily. How?

a $3\frac{2}{5} \div 3\frac{2}{5}$ b $1\frac{1}{2} \times \frac{2}{3} \div \frac{2}{3}$ c $1\frac{1}{2} \times \frac{2}{3} \times \frac{3}{2}$

C

4 Work out:

a $\frac{4}{5} \times \frac{3}{4}$ e $2\frac{1}{2} \times 6$ i $6\frac{1}{2} \times \frac{2}{5}$

b $\frac{4}{5} \div \frac{3}{4}$ f $2\frac{1}{2} \div 5$ j $\frac{3}{4} \div 3\frac{1}{2}$

c $\frac{5}{8} \times \frac{1}{3}$ g $10 \times 2\frac{3}{5}$ k $3\frac{3}{4} \times 2\frac{1}{2}$

d $\frac{5}{8} \div \frac{1}{3}$ h $12 \div 1\frac{3}{4}$ l $2\frac{3}{5} \div 2\frac{1}{6}$

5 a   How many times does $3\frac{1}{2}$ go into 21?

b   How many times does 21 go into $3\frac{1}{2}$?

6 Work out:

a $\left(1\frac{1}{4}\right)^2$ b $\frac{3}{4} + 2\frac{1}{2} \times 1\frac{1}{2}$ c $\left(\frac{3}{4} + 2\frac{1}{2}\right) \times 1\frac{1}{2}$

7 a   Which is bigger, $\frac{5}{6} \times \frac{2}{7}$ or $\frac{5}{6} + \frac{2}{7}$?
Explain how you got your answer.

b   Which is bigger, $2\frac{1}{2} \times 1\frac{1}{4}$ or $2\frac{1}{2} + 1\frac{1}{4}$?
Explain how you got your answer.

8 Jack has a dog that eats $\frac{2}{3}$ of a tin of food each day.
How many tins will be needed to feed the dog for five days?

9 A recipe for 16 biscuits needs $\frac{2}{3}$ of a cup of flour.
How much flour is needed for 48 biscuits?

10 Sanjay needs four curtains. Each curtain uses $3\frac{3}{4}$ yards of fabric.
How much fabric is that altogether?

11 A kilometre is about $\frac{5}{8}$ of a mile. How many miles are there in 16 km?
How many km in 100 miles?

12 Find the missing numbers. (These may be fractions or mixed numbers.)

a $\frac{1}{2} \times \square = 1$ c $2\frac{1}{2} \times \square = 4$

b $\frac{1}{2} \div \square = 1$ d $\square \div \frac{3}{4} = 6$

Learn... 3.5 Fractions of quantities

How do you work out three-quarters of something?

Without a calculator, a good way is to work out $\frac{1}{4}$ first, then find $\frac{3}{4}$.

Example: Find $\frac{3}{4}$ of £100.

Solution: $\frac{1}{4}$ of £100 = £100 ÷ 4 = £25

So $\frac{3}{4}$ of £100 = 3 × £25 = £75

Example: Five people have a meal and share the bill of £77.55 equally between them.

How much does each person pay?

Solution: The amount each person has to pay is one-fifth of the whole bill. To find one-fifth of something, divide it by 5.

$\frac{1}{5}$ of £77.55 = £77.55 ÷ 5 = £15.51 $\quad \begin{array}{r} 15.51 \\ 5)\overline{77.55} \end{array}$

Example: Suppose one of the people in the previous example pays for his friend as well. How much does he pay?

Solution: He pays $\frac{2}{5}$ of £77.55.

$\frac{1}{5}$ of £77.55 = £77.55 ÷ 5 = £15.51

So $\frac{2}{5}$ of £77.55 = 2 × £15.51 = £31.02

$$\begin{array}{r} 15.51 \\ \times \quad\quad 2 \\ \hline 3_11.02 \end{array}$$

To find a fraction of a quantity, divide the quantity by the denominator of the fraction, then multiply the result by the numerator of the fraction. (Alternatively you can multiply the quantity by the numerator and divide by the denominator).

Hint

Another way is to multiply by 2, then divide by 5:

£77.55 × 2 = £155.10

£155.10 ÷ 5 = £31.02

This gives the same answer, so it does not matter whether you divide by 5 first and then multiply by 2, or multiply by 2 first, then divide by 5.

AQA *Examiner's tip*

This is a non-calculator unit, so you must know how to multiply and divide without a calculator.

Example: Find $\frac{5}{6}$ of 216.

Solution: $\frac{5}{6}$ of 216 = $\frac{216}{6}$ × 5 = 36 × 5 = 180

AQA *Examiner's tip*

This is an important process – make sure you really understand it and can use it in different circumstances, with and without a calculator.

Practise... 3.5 **Fractions of quantities** 🔑

G F E D C

F

1 Work out:

    a one-eighth of 32   **c** one-tenth of 340

    b one-sixth of 24   **d** one-hundredth of £15

2 **a** Find $\frac{2}{5}$ of these numbers.

    i 20   **ii** 25   **iii** 30   **iv** 35

Explain why the answers go up by two each time.

By how much would the answers go up each time if you calculated $\frac{4}{5}$ of the numbers instead of $\frac{2}{5}$?

3 Find:

    a three-eighths of £200

    b five-twelfths of 1200 kg

    c four-ninths of £180

    d six-elevenths of 33 km

4 Find three-quarters of these quantities.

    a 100 m   **c** 10 km   **e** 20 cm^2

    b 800 g   **d** 1 metre

5 Which is bigger, $\frac{4}{5}$ of 1 kg or $\frac{3}{4}$ of 1.2 kg?

6 Amy has £120. She gives one-quarter of it to her brother and two-thirds of the remainder to her sister. She spends £10 on a DVD.
How much money does Amy have left?

7 Three-fifths of a number is 27.
What is the number?

8 A pint is 20 fluid ounces. Chris uses $\frac{3}{4}$ of a pint of milk in a recipe.
How many fluid ounces does she use?

9 A college has a grant of £9000 to spend on ICT equipment.
Two-thirds of it is to be spent on laptops and one-third on software.
How much is spent on laptops? How much is spent on software?

10 **a** Work out what these prices are when reduced in a sale.

        i A dress costing £48 is reduced by one-third.

        ii A refrigerator costing £325 is reduced by a fifth.

        iii A CD player costing £55 is reduced by a quarter.

        iv A television costing £325.40 is reduced by a tenth.

    b In a sale the price of a coat is reduced by one-fifth to £80.
What was the original price?

11 Three-quarters of one number is the same as half of another number.
What could the two numbers be?

F

Learn... 3.6 One quantity as a fraction of another

One of the most useful fraction calculations is to work out one number or quantity as a fraction of another. In real life, it is usual to turn the fractions into decimals or percentages so that they can be easily compared.

To work out one quantity as a fraction of another, change both quantities to the same units if necessary. Write the first quantity as the numerator and the second quantity as the denominator and simplify the fraction.

Example: What fraction of £5 is 25p?

Solution: First change £5 to 500p.

$$\frac{25}{500} = \frac{1}{20}$$

÷ 25

÷ 25

Simplify the fraction by dividing the numerator and the denominator by the common factor 25.

> **AQA Examiner's tip**
>
> Make sure you divide the numerator by the denominator, not the other way round.

The fraction in its simplified form is $\frac{1}{20}$.

Example: Some patients are taking part in a medical trial. 150 of them with a disease are given Drug A and 102 of them get better. 120 of them are given Drug B and 80 of them get better.

Find the fraction of patients who get better with each drug and simplify the fractions.

Solution: The fraction of patients who get better with Drug A is

$$\frac{102}{150} = \frac{51}{75} = \frac{17}{25}$$

÷ 2   ÷ 3

÷ 2   ÷ 3

> **AQA Examiner's tip**
>
> Simplify your fractions whenever possible.

The fraction of patients who get better with Drug B is

$$\frac{80}{120} = \frac{8}{12} = \frac{2}{3}$$

÷ 10   ÷ 4

÷ 10   ÷ 4

(Note: in real life, doctors need to compare the fractions to see which drug seems to be more effective. They would change the fractions to decimals or percentages.)

Example: Three-tenths of a number is 90. What is the number?

Solution: Three-tenths of the number is 90, so one-tenth of the number is 90 ÷ 3 = 30

So the whole of the number (ten tenths) is 30 × 10 = 300

Practise... 3.6 One quantity as a fraction of another

G F E D C

D

1 Work out the first number or quantity as a fraction of the second.

   a 150, 250   **c** 800 g, 2 kg

   b 50p, £4.50   **d** 75 cm, 120 m

2 At a football match, the crowd was 15 000. There were 10 500 home supporters. What fraction of the crowd was this?

3 Clare makes a cheesecake. She has 500 g of cream cheese and uses 350 g of it. What fraction of the cream cheese does she use?

D

4 Sue has 20 shirts to iron. What fraction of them has she done when she has ironed 12?

5 Mr Howes is marking 35 books. What fraction does he still have left to do when he has marked 14 books?

6 Sara's mark in one spelling test is 15 out of 20 and in the next is 20 out of 25. Which mark was better?

7 18 out of 20 in class 9Y passed a maths test and 25 out of 30 in class 9X passed. Which class did better?

8 In Kate's house, 16 of her 20 lightbulbs are low-energy ones.
In Jane's house, 20 out of 24 bulbs are low-energy.

a Who has the higher fraction of low-energy light bulbs?

b There are 30 bulbs in Dipak's house. How many low-energy bulbs does Dipak need so that he has at least the same fraction as Jane?

3 Assess 🄺

1 Write down what fraction of each shape is shaded.

G

a

b

2 Simplify these fractions.

F

a $\frac{25}{100}$ **c** $\frac{75}{100}$ **e** $\frac{15}{45}$ **g** $\frac{18}{81}$

b $\frac{50}{100}$ **d** $\frac{100}{100}$ **f** $\frac{56}{72}$ **h** $\frac{12}{60}$

3 Find three-quarters of:

a £3   **c** 1.8 kg   **e** 0.5 metres

b $98   **d** £1024

4 Find the calculations that give five-sixths of 48:

a $48 \div 6 \times 5$ **c** $48 \times 6 \div 5$ **e** $5 \times 48 \div 6$

b $48 \times 5 \div 6$ **d** $5 \div 6 \times 48$ **f** $\frac{48}{6} \times 5$

5 Four-fifths of a number is 88. What is the number?

E

6 Which fraction in this list is the greatest?

$\frac{2}{3}, \frac{7}{10}, \frac{3}{5}, \frac{8}{15}$

E **7** Write these fractions as decimals.

    a $\frac{3}{4}$ **b** $\frac{3}{5}$ **c** $\frac{7}{10}$ **d** $\frac{73}{100}$ **e** $\frac{5}{8}$

D **8** Work out the first quantity as a fraction of the second.

    a 24, 36   **c** 10 cm, 1 m

    b £2, £4.50   **d** 150 g, 1.5 kg

C **9** Lawn turf costs £12 a square metre.
What does a rectangle of lawn turf measuring $\frac{3}{4}$ metre by $\frac{5}{6}$ metre cost?

10 Sue makes trousers for her twin toddlers. Each pair of trousers needs three-eighths of a yard of fabric.
How much fabric is needed for four pairs of trousers?

11 Ali needs $1\frac{1}{3}$ cups of sugar to make fudge and $\frac{3}{4}$ of a cup of sugar to make biscuits. He has only 2 cups of sugar. How much more does he need?

12 How much bigger is $2\frac{3}{5} \times 1\frac{1}{2}$ than $2\frac{3}{5} \div 1\frac{1}{2}$?
Show how you worked it out.

AQA Examination-style questions

1 There are 24 passengers on a bus.
$\frac{1}{4}$ of the passengers are men.
$\frac{1}{3}$ of the passengers are women.
The rest of the passengers are children.

How many passengers are children?   *(3 marks)*

AQA 2007

4 Decimals

Examiners would normally expect students who get these grades to be able to:

G

round to the nearest integer

F

write down the place value of a decimal digit such as the value of 3 in 0.63

order decimals to find the biggest and the smallest

round numbers to given powers of 10 and up to 3 decimal places

E

round a number to one significant figure

add and subtract decimals

estimate answers to calculations involving decimals

D

multiply decimals such as 2.4×0.7

convert simple fractions to decimals and decimals to fractions

C

divide a number by a decimal such as $1 \div 0.2$ and $2.8 \div 0.7$

recognise that recurring decimals are exact fractions and that some exact fractions are recurring decimals.

Did you know?

Photo finish

In many sports, gold medals are won by fractions of a second.
In the 100 m sprint, every hundredth of a second counts. Since 1975, official races have been timed electronically to a hundredth of a second.
In 2009, Jamaican athlete Usain Bolt set the world record for the 100 m at 9.58 seconds. This was an improvement of 0.11 seconds from his previous record a year earlier. This doesn't seem like much, but over 100 metres it really makes a difference!

Key terms

decimal	round
integer	decimal place
place value	numerator
digit	denominator
significant figure	recurring decimal

You should already know:

✓ how to arrange whole numbers in order of size

✓ how to add, subtract, multiply and divide whole numbers.

Learn... 4.1 Place value

The **decimal** point separates the whole number or **integer** part from the fraction part.

For example, the number 318.96 can be written in a **place value** table like this:

Thousands	Hundreds	Tens	Units	.	Tenths	Hundredths	Thousandths
	3	1	8	.	9	6	

The value of the **digit** 3 is 300. The digit 3 has the highest place value so is the most important part of the number. It is called the most **significant figure.**

The value of the digit 1 is 10.

The value of the digit 8 is 8.

The value of the digit 9 is 0.9

The value of the digit 6 is 0.06

Example: Write these numbers in order of size, starting with the highest.

  17.6 16.68 17.06 17.638

Solution: The numbers should be put into a place value table.

Thousands	Hundreds	Tens	Units	.	Tenths	Hundredths	Thousandths
		1	7	.	6		
		1	6	.	6	8	
		1	7	.	0	6	
		1	7	.	6	3	8

Compare the most significant figures first. In this case these are the tens.

All four numbers start with 1 ten, so you cannot order using the tens.

Compare the units, as these are the next most significant figures.

The highest are the three numbers with 7 units, so compare the tenths for these three numbers.

Thousands	Hundreds	Tens	Units	.	Tenths	Hundredths	Thousandths
		1	7	.	6		
		1	7	.	0	6	
		1	7	.	6	3	8

17.6 and 17.638 are higher than 17.06, as 6 tenths is higher than 0 tenths.

Compare the hundredths for 17.6 and 17.638

17.6 can be written as 17.60, so 17.638 is higher.

So this gives the order:

17.638 17.6 17.06 16.68

AQA *Examiner's tip*

Make sure you read carefully which order you are asked for. In this case you were asked to start with the highest.

Practise... **4.1 Place value**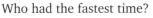

1 Write these numbers in a place value table.

    a 137.435   **c** 75.9   **e** 38.49

    b 36.401   **d** 0.72

2 Put each list of numbers in order of size, starting with the highest.

    a 5.125   5.16   5.2   5.19   5.08

    b 36.2   37.68   36.34   36.02   35.75

    c 0.096   0.35   0.46   0.64   0.421

3 Put each list of numbers in order of size, starting with the lowest.

    a 11.4   11.37   11.138   11.09   11.2

    b 5.46   6.44   5.49   7.03   5.07

    c 0.37   0.73   0.378   0.345   0.718

4 Write down the value of the digit 2 in each of these numbers.

    a 2.7   **c** 1.237   **e** 216.4   **g** 723.46

    b 7.42   **d** 21.714   **f** 0.172   **h** 7432.1

5 Here are the times, in seconds, of five runners in a 100 m race.

        Tomas 11.06

        Elijah 11.18

        Bradley 11.44

        James 11.47

        Deji 11.22

    Who had the fastest time?

Learn... **4.2 Rounding**

It is often sensible to **round** figures to give an approximate answer.

For example, using a calculator, an area is worked out to be 18.27146 square metres.

This could be rounded to 18 square metres to make the numbers more manageable.

Numbers can be rounded to the nearest integer (whole number), nearest 10, nearest 100, etc.

Numbers can also be rounded to decimals, for example one **decimal place**, depending on what the information is needed for.

Sometimes a number is exactly halfway between two others.

In this case it is rounded up to the higher number.

So 17.5 would round up to 18.

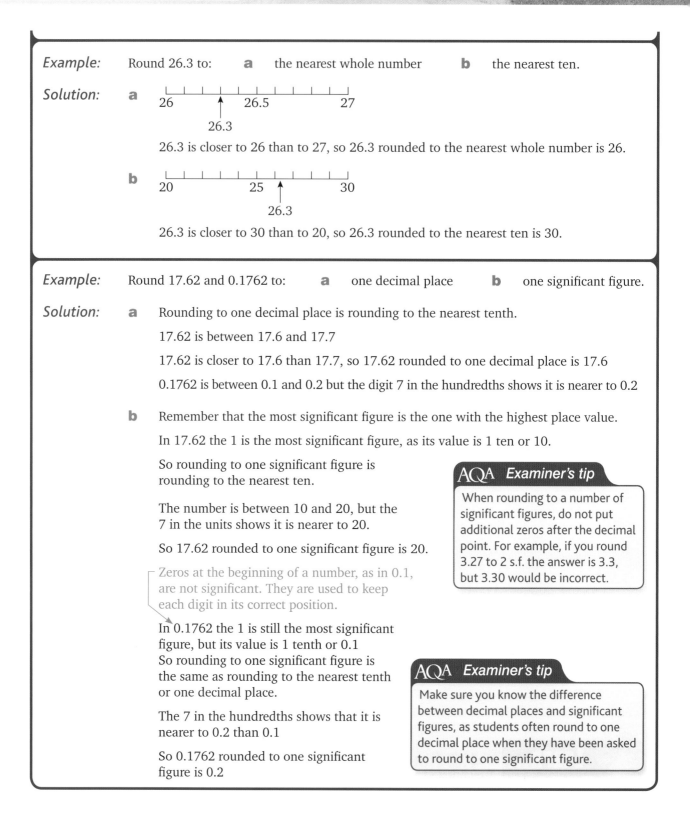

Example: Round 26.3 to: **a** the nearest whole number **b** the nearest ten.

Solution: **a**

26.3 is closer to 26 than to 27, so 26.3 rounded to the nearest whole number is 26.

b

26.3 is closer to 30 than to 20, so 26.3 rounded to the nearest ten is 30.

Example: Round 17.62 and 0.1762 to: **a** one decimal place **b** one significant figure.

Solution: **a** Rounding to one decimal place is rounding to the nearest tenth.

17.62 is between 17.6 and 17.7

17.62 is closer to 17.6 than 17.7, so 17.62 rounded to one decimal place is 17.6

0.1762 is between 0.1 and 0.2 but the digit 7 in the hundredths shows it is nearer to 0.2

b Remember that the most significant figure is the one with the highest place value.

In 17.62 the 1 is the most significant figure, as its value is 1 ten or 10.

So rounding to one significant figure is rounding to the nearest ten.

The number is between 10 and 20, but the 7 in the units shows it is nearer to 20.

So 17.62 rounded to one significant figure is 20.

Zeros at the beginning of a number, as in 0.1, are not significant. They are used to keep each digit in its correct position.

In 0.1762 the 1 is still the most significant figure, but its value is 1 tenth or 0.1
So rounding to one significant figure is the same as rounding to the nearest tenth or one decimal place.

The 7 in the hundredths shows that it is nearer to 0.2 than 0.1

So 0.1762 rounded to one significant figure is 0.2

> **AQA Examiner's tip**
> When rounding to a number of significant figures, do not put additional zeros after the decimal point. For example, if you round 3.27 to 2 s.f. the answer is 3.3, but 3.30 would be incorrect.

> **AQA Examiner's tip**
> Make sure you know the difference between decimal places and significant figures, as students often round to one decimal place when they have been asked to round to one significant figure.

Practise... 4.2 Rounding 🔑 G F E D C

1 Round these to the nearest whole number. (Use a number line to help you.)

a	6.7	**c**	57.2	**e**	17.49
b	3.4	**d**	0.4	**f**	56.5

2 Round these numbers: **a** to the nearest 10 **b** to the nearest 100.

i	726	**iii**	4278	**v**	11.4
ii	371	**iv**	3252	**vi**	727

F

3 **a** Julian says that 628 rounded to the nearest 10 is 63.
What has he done wrong?

b Ravi says that 6.5 rounded to the nearest whole number is 6.
Is he correct? Give a reason for your answer.

4 Round these numbers to one decimal place.

a	3.72	**e**	8.729	**i**	49.96
b	4.27	**f**	6.364	**j**	0.17832
c	8.45	**g**	0.754	**k**	0.092
d	9.02	**h**	230.31	**l**	23.358

5 Round these numbers to: **a** two decimal places **b** three decimal places.

i	74.2387	**iii**	0.0078	**v**	0.02375
ii	0.5462	**iv**	6.0552	**vi**	19.3476

6 Round these numbers to one significant figure.

E

a	127 cm	**f**	53 782 m
b	284 kg	**g**	48 127 miles
c	3756 g	**h**	109.347
d	8429 m	**i**	0.632
e	62.7 mm	**j**	0.77214

7 Write down a number that when rounded to the nearest hundred gives the same answer as when rounded to the nearest thousand.

8 An amount of money correct to the nearest pound is £36.
Write down the smallest and largest amount it could be.

9 The number of spectators at a football match is 42 488.
A local newspaper reported this as 43 000 to the nearest thousand.
Is this correct?

10 Find a number that is the same when rounded to the nearest whole number as when rounded to one significant figure.

11 Dan rounds a number to the nearest 10 and gets 20. Megan rounds the same number to the nearest whole number and gets 15.

a What could the number be?

b How many examples can you think of?

c Does your number have to be a decimal?

12 At Leeds Rhinos' last Super League game the attendance quoted was 26 000.
What do you think this has been rounded to? How many could have been there?

13 Chloe had ten paperback books which had a width of 3 cm each to the nearest whole number.
She tried to put them on a shelf which had a width of 30 cm but they would not fit.
Why not?
Give a reason for your answer.

Learn... 4.3 Adding and subtracting decimals

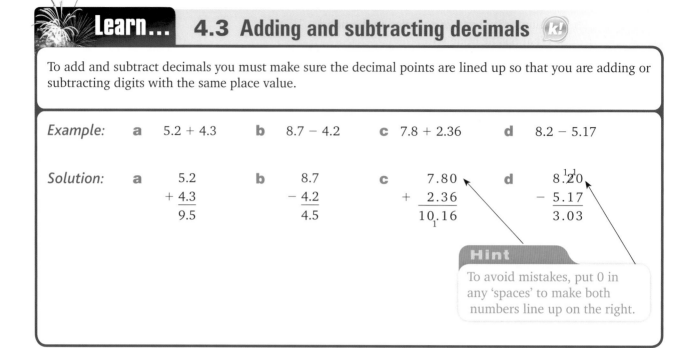

To add and subtract decimals you must make sure the decimal points are lined up so that you are adding or subtracting digits with the same place value.

Example: **a** 5.2 + 4.3 **b** 8.7 − 4.2 **c** 7.8 + 2.36 **d** 8.2 − 5.17

Solution: **a**
$$\begin{array}{r} 5.2 \\ +\ 4.3 \\ \hline 9.5 \end{array}$$
b
$$\begin{array}{r} 8.7 \\ -\ 4.2 \\ \hline 4.5 \end{array}$$
c
$$\begin{array}{r} 7.80 \\ +\ 2.36 \\ \hline 10.16 \end{array}$$
d
$$\begin{array}{r} 8.\overset{1}{2}\overset{1}{0} \\ -\ 5.17 \\ \hline 3.03 \end{array}$$

Hint

To avoid mistakes, put 0 in any 'spaces' to make both numbers line up on the right.

Practise... 4.3 Adding and subtracting decimals

G F E D C

E

1 Work out:

a 14.3 + 7.6   **g** 102.3 + 97.8

b 5.4 + 7.29   **h** 69.46 − 22.7

c 6.18 + 4.93   **i** 9.28 − 0.16

d 17.6 − 8.54   **j** 17.8 − 8.76

e 24.09 − 15.6   **k** 0.78 + 9.32 − 6.1

f 7.92 − 5.48 + 12.26   **l** 87.4 − 31.6 + 42.9

2 Greg says that 7.63 − 4.2 = 3.61

He is **not** correct.

Explain the mistake he has made and work out the correct answer.

3 Fill in the missing numbers (shown as ☺) in these calculations.

a 6.5 + 1.3 = ☺.8   **d** 3.7 − 1.2 = ☺.5

b 1.☺ + 6.2 = 7.8   **e** ☺.4 − 3.8 = 2.☺

c ☺.2 − 5.1 = 3.1   **f** 9.☺ − 4.2 = ☺.5

Hint

The missing numbers are not all the same number.

4 Meg says that 11.7 + 4.62 = 15.132

Tina says that 11.7 + 4.62 = 16.32

Liz says that 11.7 + 4.62 = 15.9

Who is correct?
Explain the mistakes the other two made.

5 A college snack bar has the following menu.

Food

Sandwich £1.65

Toasted sandwich add 30p

Panini £2.45

Jacket potato and cold filling £1.80

Jacket potato and hot filling £2.00

Soup £1.25

Drinks

Tea or coffee £1.05

Cold drink £0.95

a Carl buys soup, a sandwich and a cold drink.
How much does he pay?

b Jill buys a jacket potato and cold filling and a coffee.
She pays with a £5 note.
How much change should she receive?

c Zac wants something to eat and something to drink
but only has £2.75.
List the combinations he could afford to buy.

6 Mark is repairing some floorboards in his house.
He wants to replace four pieces of floorboard.
He needs two pieces of length 1.5 metres, one of length 0.62 metres and one of
length 0.38 metres.

The Do-It-Yourself store sells the following lengths of floorboards.

1.8 metres costs £3.76.

2.4 metres costs £5.20.

Mark can cut these lengths to the sizes needed for his
floorboards.

Work out the best way for Mark to buy enough floorboards
to do his repairs.

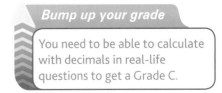

Bump up your grade

You need to be able to calculate
with decimals in real-life
questions to get a Grade C.

Learn... 4.4 Multiplying decimals

There are different ways to multiply two numbers together such as 32×17.

You can use the grid method:

×	30	2
10	300	20
7	210	14

$300 + 210 + 20 + 14 = 544$

You can use the column method:

```
    32
×   17
   224
   320
   544
```

These methods can also be used to multiply decimals.

Example: **a** Work out the exact value of 6.4×3.1

b Work out the exact value of 1.2×4

Solution: **a** First remove the decimal points: 64×31

Then multiply in your usual way.

Grid method

×	60	4
30	1800	120
1	60	4

$1800 + 120 + 60 + 4 = 1984$

Column method

```
      64
×     31
      64
    1920
    1984
```

Finally, put the decimal point back into the answer.

Estimate the answer by rounding each number to one significant figure.

So 6.4×3.1 is approximately $6 \times 3 = 18$

So the answer is 19.84
Or count up the number of decimal places in the question.

There are two decimal places in the question: 6.4×3.1

So you need two decimal places in the answer: 19.84

So $6.4 \times 3.1 = 19.84$

b Remove the decimal point: 12×4

multiply the numbers: 48

replace the decimal point by either estimating 1.2×4 is approximately $1 \times 4 = 4$

or by counting the number of decimal points in the question: one decimal point in the question, so one decimal point in the answer.

So $1.2 \times 4 = 4.8$

> **AQA** *Examiner's tip*
>
> Always estimate answers to questions involving decimals to make sure you put the decimal point in the correct place.

Practise... 4.4 Multiplying decimals (k!)

G F E D C

1 Work out:

a	2.1×4	**d**	6.3×4	**g**	0.4×8
b	5×2	**e**	0.2×6	**h**	0.6×4
c	8.2×3	**f**	0.7×5		

2 For each question, decide which is the best estimate.

		Estimate A	Estimate B	Estimate C
a	6.2×7.9	4.8	42	48
b	4.36×9.4	3.64	36	50
c	28.7×19.2	40.0	400	600

3 Work out:

a	0.3×0.3	**d**	0.15×0.3	**g**	$0.3 \times 0.2 \times 0.5$
b	2.3×0.2	**e**	0.05×0.1		
c	0.4×0.6	**f**	3.1×0.3		

4 Use the multiplication $23 \times 52 = 1196$ to help you to complete these questions.

a	2.3×52	**d**	2.3×5.2	**g**	0.23×0.052
b	0.23×52	**e**	0.23×0.52		
c	0.023×0.052	**f**	0.23×5.2		

5 Work out:

a	1.3×22	**d**	1.5×3.2	**g**	0.7×1.3
b	1.7×2.3	**e**	1.2×1.7	**h**	5.1×12.3
c	8.7×2.5	**f**	8.9×1.6		

⚠ **6**   Using your answers to question 5, write down the answers to these.

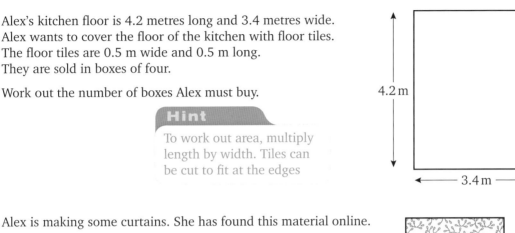

a   0.13 × 0.22   **e**   0.012 × 0.17

b   0.17 × 0.23   **f**   0.0089 × 0.016

c   0.087 × 0.025   **g**   0.07 × 1.3

d   1.5 × 0.032   **h**   0.0051 × 0.123

Bump up your grade

Bump up your grade by multiplying two decimals.

⚙ **7**   Alex's kitchen floor is 4.2 metres long and 3.4 metres wide.
Alex wants to cover the floor of the kitchen with floor tiles.
The floor tiles are 0.5 m wide and 0.5 m long.
They are sold in boxes of four.

Work out the number of boxes Alex must buy.

Hint

To work out area, multiply length by width. Tiles can be cut to fit at the edges

4.2 m

3.4 m

⚙ **8**   Alex is making some curtains. She has found this material online.

She needs four lengths of material each measuring 2.3 metres.
What is the total cost of the material she needs to buy?
Give your answer to the nearest penny.

£7.99 per metre

Learn... 4.5 Dividing decimals

To divide decimals without a calculator, you need to use numbers that are easier to work with.

First, write the division as a fraction. Then find an equivalent fraction that is easier to work with.
Multiply the numerator and denominator so that the denominator is a whole number.
You will need to multiply by 10 if there is one decimal place.
Then divide the numerator by the denominator to get your answer.

AQA *Examiner's tip*

Never try to divide by a decimal. **Always** make your divisor a whole number using equivalent fractions.

Example: **a**   Work out 26.4 ÷ 0.4   **b**   Work out 3.8 ÷ 0.02

Solution: **a**   First write the division as a fraction: $\frac{26.4}{0.4}$

Next, multiply the **numerator** and **denominator** by 10 so that the denominator is a whole number.

$$\frac{26.4}{0.4} = \frac{264}{4}$$

×10

×10

Finally divide the numerator by the denominator.

$$\begin{array}{r} 66 \\ 4\overline{)264} \end{array}$$ So 26.4 ÷ 0.4 = 66

b $3.8 \div 0.02 = \frac{3.8}{0.02}$

This time you need to multiply by 100 so that the denominator is a whole number.

$$\frac{3.8}{0.02} = \frac{380}{2} = 190$$

×100

×100

Practise... 4.5 Dividing decimals 🔊   **G F E D C**

D

1   Work out:

    a $24 \div 0.4$ **c** $81 \div 0.3$ **e** $0.8 \div 0.2$ **g** $48 \div 0.8$

    b $36 \div 0.2$ **d** $63 \div 0.7$ **f** $70 \div 0.5$ **h** $84 \div 0.4$

C

2   Work out:

    a $3.2 \div 0.4$ **d** $25.4 \div 0.2$ **g** $4.07 \div 1.1$ **j** $25.3 \div 0.11$

    b $53.1 \div 0.3$ **e** $1.74 \div 0.6$ **h** $16.8 \div 0.12$

    c $0.56 \div 0.7$ **f** $1.32 \div 0.04$ **i** $22.8 \div 1.2$

3   Carrie knows that $3.4 \div 0.4 = 8.5$

    Use this fact to copy and fill in the gaps in these questions.

    a $34 \div 0.4 = \square$ **c** $\square \div 0.04 = 8.5$ **e** $\square \div 4 = 0.85$

    b $340 \div \square = 8.5$ **d** $3.4 \div 4 = \square$ **f** $0.34 \div 8.5 = \square$

4   Stu says that $36 \div 2 = 18$, so $36 \div 0.2 = 1.8$
    Dean says $36 \div 2 = 18$, so $3.6 \div 0.2 = 1.8$
    Jack says $36 \div 2 = 18$, so $3.6 \div 2 = 1.8$
    Who is right? Give a reason for your answer.

5   Estimate the answers to each of these questions.

    a $\dfrac{78.2 \times 2.8}{0.23}$ **c** $\dfrac{26.2 + 23.9}{0.89 - 0.72}$

    b $\dfrac{107.6 + 92.1}{0.37}$ **d** $\dfrac{56.3 \times 9.7}{0.316}$

> **AQA** *Examiner's tip*
> When estimating, always round numbers to one significant figure first.

6   Jenny needs some pieces of ribbon of length 0.6 metres.
    She has a roll of ribbon of length 4.3 metres.
    How many pieces of ribbon can she cut from the roll?

7   A factory making 'Fizzypop' drink makes up batches of 2500 litres of 'Fizzypop' at a time.
    A machine fills cans with the drink.
    Each can holds 0.3 litres.
    How many cans will be filled from one batch of 'Fizzypop'?

8   Loren is going on holiday to France for five days and five nights.
    €1 is worth £0.80. Loren changes £400 into euros.
    Her hotel costs €39 per night.
    How many euros does she have left to spend per day?

9   Jamie says that when you divide one number by another the answer is always smaller.
    Give an example to show that Jamie is wrong.

Learn... 4.6 Fractions and decimals

To change a fraction to a decimal, divide the numerator (top number) by the denominator (bottom number).

$\dfrac{2}{5}$ ←—— Numerator
←—— Denominator

Some fractions become **recurring decimals**.

This means that a number or group of numbers keeps repeating.

To change a decimal to a fraction, put the decimal into a place value table.

For example, you can write the number 0.37 in a place value table like this:

All fractions can be changed to a decimal. If the decimal keeps repeating it is a recurring decimal. If it does not, it is called a **terminating** decimal.

Units	.	Tenths	Hundredths
0	.	3	7

The least significant figure is hundredths, so this is the denominator of the fraction.

So $0.37 = \dfrac{37}{100}$

Example: **a** Write $\frac{2}{5}$ as a decimal. Is the answer a recurring decimal or a terminating decimal?

b Write $\frac{1}{3}$ as a decimal. Is the answer a recurring decimal or a terminating decimal?

Solution: **a** Divide 2 by 5: $5\overline{)2.0}$ with 0.4 above

So $\frac{2}{5} = 0.4$

Add zeros to the calculation. Make sure the decimal points are lined up.

The decimal does not repeat, so this is a terminating decimal.

b Divide 1 by 3: $3\overline{)1.0000}$ with 0.3333 above the '3' keeps repeating, so the answer is a recurring decimal. This can be written as $0.\dot{3}$ (the dot shows which number is recurring).

Example: Write 0.6 as a fraction in its simplest form.

Solution:

Units	.	Tenths
0	.	6

The least significant figure is tenths, so this is the denominator of the fraction.

$0.6 = \dfrac{6}{10}$

This can be cancelled down to a simpler form by dividing both numbers by 2.

$\dfrac{6}{10} = \dfrac{3}{5}$ so $0.6 = \dfrac{3}{5}$

Practise... 4.6 Fractions and decimals (k!) G F E D C

1 Change these fractions to decimals.

a $\frac{1}{5}$ **d** $\frac{3}{20}$ **g** $\frac{1}{50}$

b $\frac{7}{10}$ **e** $\frac{3}{4}$ **h** $\frac{3}{10}$

c $\frac{1}{8}$ **f** $\frac{3}{100}$

2 Which of these fractions is closest to 0.67?

a $\frac{3}{4}$ **b** $\frac{5}{8}$ **c** $\frac{3}{5}$

D

D

3 Write these decimals as fractions.

| a | 0.59 | c | 0.4 | e | 0.1 | g | 0.45 |
| b | 0.07 | d | 0.25 | f | 0.36 | h | 0.05 |

C

4 Write these fractions as recurring decimals.

a $\frac{1}{9}$ b $\frac{2}{3}$ c $\frac{1}{6}$

↓ 5 Write these fractions as recurring decimals.

a $\frac{2}{9}$ c $\frac{5}{6}$

b $\frac{4}{15}$ d $\frac{1}{22}$

Hint

If there is more than one number in the recurring pattern, put a dot over the first and last numbers of the pattern.

? **6** On her birthday, Bridget is given a big box of small sweets called Little Diamonds.
She wants to find out how many sweets are in the box, but it would take too long to count them.
A label on the box tells her that the total weight is 500g.
She weighs 10 sweets. The weight of the 10 sweets is 0.4g.
How many sweets are there in the box?

? **7** Three identical blocks of wood are placed as shown, so that the top one rests with $\frac{1}{3}$ of its length on each of the other two.

0.42 m

Work out the length of one of the blocks.

4 Assess ⓚ

G

1 Round each number to the nearest integer.

| a | 37.2 | c | 4.295 | e | 13.526 |
| b | 9.7 | d | 3.71 | | |

F

2 Put each list of numbers in order of size, starting with the smallest.

a 7.2   7.02   7.16   7.28   7.025

b 84.72   83.9   83.531   84.709   84.8

c 0.46   0.64   0.446   0.464   0.466

3 Write down the value of the digit 8 in each of these numbers.

| a | 4.08 | c | 8.237 | e | 4.008 | g | 836.9 |
| b | 7.84 | d | 86.7 | f | 0.86 | h | 8342.1 |

4 32 917 people attend a concert. Write this number:

a to the nearest 1000

b to the nearest 10

c to the nearest 10 000

d to the nearest 100.

5 Ali drives 12.4 miles to get to work each day.
Petra drives 3.7 miles further than Ali to get to work.
How far does Petra drive?

6 Jon says that he and Carl are the same weight – to one significant figure.
Jon weighs 76 kilograms.
Carl weighs 82 kilograms.
Is Jon correct?
Explain your answer.

7 In a store room there are two piles of magazines.
One of the piles is 30 cm high.
The other pile is 12 cm high.
Each magazine is 0.6 cm thick.
How many magazines are in the store room?

8 Tom buys his electricity from Lowlec.
Lowlec charge the following prices:

Daytime electricity £0.20 per unit

Night-time electricity £0.04 per unit.

From June to September Tom uses 450 units of daytime electricity.
His total bill is £122.
Work out the number of night-time electricity units Tom used during this period.

AQA Examination-style questions

1 You are making bookcases from planks of wood that are 20 cm wide and 20 mm thick.
The planks are sold in these lengths:
1.8 m   2.1 m   2.4 m   2.7 m   3 m

You make this bookcase.

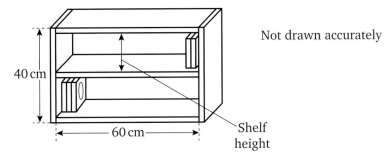

Not drawn accurately

40 cm

60 cm

Shelf height

   a What is the total length of wood used to make the bookcase? *(2 marks)*

   b The shelf height is the height of the gap between the shelves.
The two shelf heights are equal.
Work out the shelf height. *(3 marks)*

AQA 2009

2 Sam has £1.65
Vicki has 7p
How much must Sam give Vicki so that they each end up with the same amount? *(3 marks)*

AQA 2008

5 Working with symbols

Objectives

Examiners would normally expect students who get these grades to be able to:

F

simplify an expression such as $5a + 2a - 3a$

work out the value of an expression such as $3x + 2y$ when $x = 4$ and $y = 3$

E

simplify an expression such as $4a + 5b - a + 2b$

understand the rules of arithmetic as applied to algebra, such as $x - y$ is not equal to $y - x$

work out the value of an expression such as $2x - 3y$ for negative values of x and/or y

D

expand brackets such as $4(x - 3)$

factorise an expression such as $6x + 8$

C

expand and simplify an expression such as $3(3x - 7) - 2(3x + 1)$

Did you know?

$$E = mc^2$$

This is probably the most famous piece of algebra ever written down. It is a formula written by the scientist Albert Einstein in 1905 to explain the relationship between matter and energy.

In April 2008, the American singer Mariah Carey had a number one hit in the United States with her album called $E = mc^2$

Key terms

simplify	unlike terms
expression	substitution
term	expand
like terms	factorise

You should already know:

✔ number operations and BIDMAS

✔ how to add and subtract negative numbers

✔ how to multiply and divide negative numbers

✔ how to find common factors.

Learn... 5.1 Introducing symbols and collecting like terms

What do we mean by $5a + 2a - 4a$?

Here the letter symbol a represents an unknown number, so $5a$ means five lots of that number, $2a$ means two lots of that number and $4a$ means four lots of that number.

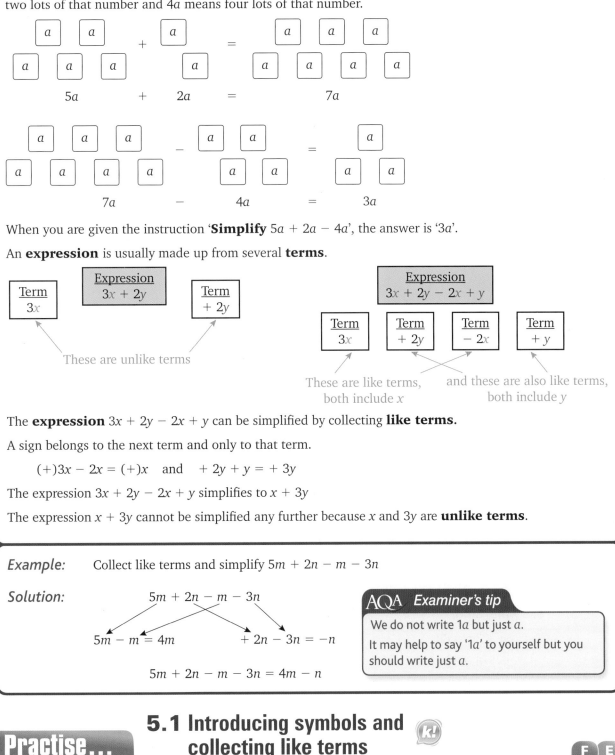

When you are given the instruction '**Simplify** $5a + 2a - 4a$', the answer is '$3a$'.

An **expression** is usually made up from several **terms**.

These are unlike terms

These are like terms, both include x

and these are also like terms, both include y

The **expression** $3x + 2y - 2x + y$ can be simplified by collecting **like terms**.

A sign belongs to the next term and only to that term.

$$(+)3x - 2x = (+)x \quad \text{and} \quad + 2y + y = + 3y$$

The expression $3x + 2y - 2x + y$ simplifies to $x + 3y$

The expression $x + 3y$ cannot be simplified any further because x and $3y$ are **unlike terms**.

Example: Collect like terms and simplify $5m + 2n - m - 3n$

Solution:
$$5m + 2n - m - 3n$$
$$5m - m = 4m \qquad + 2n - 3n = -n$$
$$5m + 2n - m - 3n = 4m - n$$

AQA *Examiner's tip*

We do not write $1a$ but just a.

It may help to say '$1a$' to yourself but you should write just a.

5.1 Introducing symbols and collecting like terms

Practise...

F E

1 Simplify:

a $a + a + a + a$	**e** $2y + y + 7y$
b $b + b + b - b - b$	**f** $4z + 5z + 6z$
c $c - c + c - c$	**g** $a + 6a - 4a$
d $3x + 4x + 5x$	**h** $2b + 8b - 3b$

i $12c - 5c + 4c$

j $4p + 5p - p - 2p$

k $11q - 3q - 4q + 5q$

l $5t - 2t + 3t - 6t$

F

F

2 Jack simplifies $7x + 3x - 2x - x$

He writes down 7 as his answer.

Explain his mistake.

3 Simplify:

a $5a + a - 8a$ **b** $3b - 5b - 2b$

E

4 Collect like terms and simplify:

a $p + p + r + p$ **g** $x + 6y - 4y + 3x$

b $x + y + y - x$ **h** $2p + 5r + r - p$

c $k - m - k + m + k$ **i** $8s + t - 3s + 2t$

d $3a + 7b + a + 2b$ **j** $3v - 3w - 2v + 4w$

e $5c + 3d - 2c - 2d$ **k** $5x - y - 2x - 6y$

f $9e + 2f - 5e + 3f$ **l** $z + 3u - 3z - u$

5 Collect like terms and simplify:

a $2m + n - 3p - m + 4n - 5p$

b $3x + 8 + 2x - 5$

c $5 + y - 3 + 6y$

d $12 - 4k + 7k - 3$

e $6t + 7 - 5z - 3z + t - 11$

6 Write an expression with four terms which simplifies to $4a + 9b$

7 Write an expression with four terms which simplifies to $8x - 5y$

8 Write an expression with six terms which simplifies to $3p - 7q - 4$

9 Kim tries to simplify $5a + 2b - 3a + b$

She writes down:

$5a + 3a = 8a$

$2b - b = 1b$

Answer $8a + 1b$

Describe her mistakes and write down the correct answer.

10 Three students are trying to simplify $6x + 7y - 2z + y + 3z - x$

Lee's answer is $6 + 8y - z$

Mark's answer is $5x + 8y - 5z$

Nat's answer is $5x + 8y + z$

Which student has the correct answer?

Describe the mistakes made by the other two students.

5.2 The rules of arithmetic and the rules of algebra

Learn...

Algebraic expressions obey the rules of arithmetic. Remember that letters represent numbers.

Addition

7 + 2 is the same as 2 + 7 $x + y$ is the same as $y + x$

Subtraction

7 − 2 is **not** the same as 2 − 7 $x − y$ is **not** the same as $y − x$

Multiplication

3 × 5 is the same as 5 × 3 $a × b$ is the same as $b × a$

Division

$\frac{3}{5}$ is **not** the same as $\frac{5}{3}$ $\frac{a}{b}$ is **not** the same as $\frac{b}{a}$

More about multiplication

$2a$ means   2 lots of a or $2 × a$

$5b$ means   5 lots of b or $5 × b$

ab means a lots of b or $a × b$

$3ab$ means 3 lots of ab or $3 × ab = 3 × a × b$

abc means a lots of bc or $a × bc = a × b × c$

Remember that $a × a$ is written as a^2, so $3a^2$ means $3 × a × a$

but $(3a)^2$ means $3a × 3a = 9a^2$

AQA *Examiner's tip*

Put the numbers first, then the letters in alphabetical order.

2b or not 2b?

4b+2b
bxb
2xb

Example: Simplify $4xy − mn + xy − 2mn$

Solution: $4xy$ and $+ xy$ are like terms, so they can be simplified to $5xy$

          $− mn$ and $− 2mn$ are like terms, so they can be simplified to $− 3mn$

          $4xy − mn + xy − 2mn = 5xy − 3mn$

Example: Simplify $a^2 + 3a^2 + a$

Solution: a^2 and $+ 3a^2$ are like terms, so they can be simplified to $4a^2$

          $4a^2$ and $+ a$ are unlike terms, so there is no more simplification.

          $a^2 + 3a^2 + a = 4a^2 + a$

5.2 The rules of arithmetic and the rules of algebra

Practise...

G F E D C

1   Write these expressions in the correct algebraic form:

a	$b × a$	**c**	$d × c$	**e**	$y × z × x$	**g**	$x × x × 4$
b	$b × a × 5$	**d**	$d × 3 × c$	**f**	$y × 8 × z × x$	**h**	$y × y × y$

2   Simplify and find the odd one out:

a $4a + 3b − 2a,$ $5b − 2a − 2b,$ $b − 3a + 2b + 5a$

b $4c + 3d − 5d + c,$ $3c + d − 8c + d,$ $2c + 6d − 7c − 4d$

c $7e − f − 3f − 6e,$ $9e − f − 2e − 3e,$ $2f − 6f − 8e + 9e$

G

E

E

3 Simplify:

a $xy + 5xy$

b $7ab - 4ab$

c $2pq - pq + 4pq$

d $3ab + 2cd + ab + 7cd$

e $2ef + 5gh - 4gh + ef$

f $5abc - 2pq - 3abc - 3pq$

g $2x^2 + 5y - 3y + x^2$

h $a^2 - 4a + 3a^2 + 2a$

i $7p^2 + 2q^2 - 3p^2 + q^2$

j $3x^2 - 2x - x^2 + x$

k $y + 4y^2 - 2y^2 + 6y$

4 Write down an expression with four terms which simplifies to $4xy + 3uv$

5 Write down an expression with four terms which simplifies to $pq - 9st$

6 Write 'true' or 'false' for each of these statements.

a $5x + y$ is always the same as $y + 5x$

b $7p - 5q$ is always the same as $5q - 7p$

c $8ef$ is always the same as $8fe$

d $\dfrac{m}{n}$ is always the same as $\dfrac{n}{m}$

e $5k^2$ is always the same as $(5k)^2$

7 x and y are integers.

Both x and y are odd numbers.

Which of these expressions are also odd?

a $x + y$ b $x - y$ c $3x$ d xy e $2y$

8 Find three pairs of like terms from the list below.

$+5ab$ $-2a^2$ $+ba$ $-4a$ $+3b$ $+3b^2$ $+a^2$ $-2a$

Learn... 5.3 Substitution

When each of the letters in an expression represents a given number, you can find the value of that expression. This is called **substitution**.

In the first example, the letters have positive values.

Example: $x = 2$ and $y = 10$

Find the value of:

a $3x + 4y$ c $2x - 3y$ e $\dfrac{y}{x}$

b $8x - y$ d $3xy$

Solution:

a $3x + 4y = (3 \times 2) + (4 \times 10)$
$= 6 + 40$
$= 46$

b $8x - y = (8 \times 2) - 10$
$= 16 - 10$
$= 6$

> **AQA Examiner's tip**
>
> It may help to put brackets in to remind yourself which operations to do first.
> BIDMAS applies to algebra as well as arithmetic.

c $2x - 3y = (2 \times 2) - (3 \times 10)$

   $= 4 - 30$

   $= -26$

d $3xy = 3 \times 2 \times 10$

   $= 60$

e $\dfrac{y}{x} = \dfrac{10}{2}$

   $= 5$

In this second example, the letters have negative values.

Example: $z = -2$ and $t = -3$

Find the value of:

a $2z + 3t$ **c** $3t - 2z$ **e** $\dfrac{9z}{t}$

b $2z - 3t$ **d** $z^2 + t^2$

Solution: **a** $2z + 3t = (2 \times -2) + (3 \times -3)$

   $= -4 + -9$

   $= -13$

Hint

$+ \times - = -$

$- \times - = +$

So the square of any number, positive or negative, is always positive.

b $2z - 3t = (2 \times -2) - (3 \times -3)$

   $= -4 - -9$

   $= -4 + 9$

   $= 5$

c $3t - 2z = (3 \times -3) - (2 \times -2)$

   $= -9 - -4$

   $= -9 + 4$

   $= -5$

d $z^2 + t^2 = (-2)^2 + (-3)^2$

   $= +4 + +9$

   $= 13$

e $\dfrac{9z}{t} = \dfrac{9 \times -2}{-3}$

   $= \dfrac{-18}{-3}$

   $= 6$

Hint

$+ \div - = -$

$- \div - = +$

Practise... 5.3 Substitution 🔊   F E D C

1 $a = 5$ and $b = 2$

Find the value of:

a $a + b$ **d** $4a - 7b$ **g** ab

b $2a + 5b$ **e** $6a - 3b$ **h** $3ab$

c $5b + 2a$ **f** $4b - a$ **i** $\dfrac{a}{b}$

AQA *Examiner's tip*

Remember $ab = a \times b$

When $a = 5$ and $b = 2$

ab is NOT equal to 52.

F

F

2 $p = 4$ and $t = 7$

Find the value of:

   a $2p + t$ **d** $3p + 4t$ **g** pt

   b $3t - p$ **e** $4p - 2t$ **h** $5pt$

   c $p + 5t$ **f** $p - t$ **i** $\dfrac{2t}{p}$

E

3 $z = 7$ and $t = 9$

Write down an expression that starts with the term $4z$ and has the value 1.

4 $x = 3$ and $y = 1$

   a Show that $2x + 4y = 10$

   b Write down two different expressions in x and y that have the value 10.

5 $c = 6$ and $d = -2$

Find the value of:

   a $c + d$ **d** $4c + 3d$ **g** cd

   b $c + 3d$ **e** $c - d$ **h** $3cd$

   c $2c + 5d$ **f** $c - 3d$ **i** $\dfrac{c}{d}$

6 $m = -3$ and $n = -5$

Find the value of:

   a $m + n$ **d** $m - n$ **g** mn

   b $2m + 3n$ **e** $m - 2n$ **h** $4mn$

   c $5m + 2n$ **f** $n - m$ **i** $\dfrac{5m}{n}$

7 $x = 8$ and $y = -1$

   a Show that $3x + 4y = 20$

   b Write down two different expressions in x and y that have the value 20.

8 $z = -3$ and $t = -4$

Write down an expression that starts with the term $2zt$ and has the value 9.

9 Zoe says that when n is an even number, $2n + 1$ is always a prime number.

   a Show that this is true when $n = 5$

   b Give an example to show that it is not always true.

10 $c = -2$ and $d = -4$

Write down an expression that starts with the term $\dfrac{6c}{d}$ and has the value 7.

11 Cara says that $2x - y$ can never be equal to $y - 2x$

Lauren says that they are equal if $x = 3$ and $y = 6$

Can you find another pair of values for which these two expressions are equal?

What is the rule for finding them?

How many pairs of values are there?

Learn... 5.4 Expanding brackets and collecting terms

When you **expand** a bracket, **all** the terms inside the bracket must be multiplied by the term outside the bracket.

You will be given the instruction **expand** or **multiply out**.

Example: **1**   Expand $3(x - 2)$

$\times$	x	-2
3	$3x$	-6

$3(x - 2) = 3x - 6$

If we know the value of x we could check that the value of the two expressions is the same.

If $x = 5$:

$3(5 - 2) = 3 \times 3$
$\qquad\qquad = 9$

and

$3 \times 5 - 6 = 15 - 6$
$\qquad\qquad = 9$

2   Multiply out $5a(2a + 1)$

$\times$	$2a$	$+1$
$5a$	$10a^2$	$+5a$

$a \times a = a^2$

$5a(2a + 1) = 10a^2 + 5a$

3   Expand and simplify $9x - 2(x - 4)$

$\times$	x	-4
-2	$-2x$	$+8$

$9x - 2(x - 4) = 9x - 2x + 8$
$\qquad\qquad\qquad = 7x + 8$

4   Expand and simplify $4(3y + 2) - 5(y - 3)$

Step 1   Expand $4(3y + 2)$

$\times$	$3y$	$+2$
4	$12y$	$+8$

$4(3y + 2) = 12y + 8$

Step 2   Expand $-5(y - 3)$

$\times$	y	-3
-5	$-5y$	$+15$

$-3 \times -5 = +15$

$-5(y - 3) = -5y + 15$

Step 3   Merge the two answers by collecting like terms

$4(3y + 2) - 5(y - 3) = \boxed{12y} + 8 \boxed{-5y} + 15$
$\qquad\qquad\qquad\qquad\quad = 7y + 23$

> **Hint**
> Underlining or circling like terms (including their signs) helps when you are collecting them.

5.4 Expanding brackets and collecting terms

Practise...

k!

G F E D C

D

1 Multiply out:

a $2(a + 3)$ e $2(3f - 2)$ i $r(5r + 4)$

b $4(b - 1)$ f $4(p + 2q)$ j $3x(2x - 3)$

c $3(2c + 3)$ g $p(p - 2)$ k $5m(1 - 2n)$

d $5(d + e)$ h $q(1 + 3q)$ l $2pq(p - 3q)$

2 The answer to an expansion is $3x + 6y$

What was the expression before it was expanded?

3 The answer to an expansion is $pq - pr$

What was the expression before it was expanded?

4 Joe says the expansion of $4ab(2c - d)$ is $8abc - d$

Kevin says it is $4abc - 4abd$

Liam's answer is $8ab - abd$

All their answers are incorrect.

Describe their mistakes and write down the correct answer.

C

5 Expand and simplify:

a $4(a + 2) + 3(a + 3)$ h $4(2r + 5) - 3(r - 2)$

b $2(b - 4) + 3(2b + 1)$ i $5(t - 3) - 3(2 - t)$

c $4(2c - 3) + 5(c - 2)$ j $6(2 - x) - (3 - x)$

d $5(p - 1) + 3(2p - 1)$ k $y(y + 5) + 2y(y - 3)$

e $5y - 3(y - 1)$ l $3k(1 + k) - k(k + 6)$

f $8 - 2(4 - 3m)$ m $3(a + b) + 2(a - b)$

g $3(4q - 1) - 2(3q + 4)$ n $4(2c - d) - 7(c - 2d)$

> **Hint**
> Think of '$- (3 - x)$' as '$- 1(3 - x)$'.

> **Bump up your grade**
> To get a Grade C you need to be able to expand and simplify in the same expression.

6 $4(x - 3) + 2(3x + 8) = ax + b$

Work out the values of a and b.

7 Each of the four cards A, B, C and D has an algebraic expression on it.

A	B	C	D
$5x + 1$	$2x - 7$	$4 - 3x$	$x + 3$

a Expand and simplify:

  i $B + 2C$

  ii $2A + 3C$

  iii $4D + B$

b Show that $B + C + D$ is equal to zero.

c Work out a combination of three of these cards that gives the answer 12.

8 Kate thinks the answer to expanding $3(q - 5) - 2(7 - q)$ is $q - 29$

What mistake did she make?

Learn... 5.5 Factorising expressions *k!*

Factorising algebraic expressions is the inverse operation of expanding. The common factor could be a number, a letter, or both.

Questions 2 and 3 in the last exercise asked you to find the expression before it was expanded, but you will usually be given the instruction **factorise**.

Example:

1 Factorise $3p + 6$

$3p + 6 = 3(p + 2)$ ⟵ $3p = 3 \times p$ and $6 = 3 \times 2$
Both terms have 3 as a common factor.

2 Factorise $q^2 - 5q$

$q^2 - 5q = q(q - 5)$ ⟵ $q^2 = q \times q$ and $5q = 5 \times q$
Both terms have q as a common factor.

3 Factorise $4xy - 2x$

$4xy - 2x = 2x(2y - 1)$ ⟵ $4xy = 2x \times 2y$ and $2x = 2x \times 1$
Both terms have $2x$ as a common factor.

Practise... 5.5 Factorising expressions *k!*   **D**

1   Match each expression with the correct factors.

Expression	Factors
$8a + 10b$	$2(4a - 5b)$
$10b - 8a$	$2(5a - 4b)$
$8a - 10b$	$2(4a + 5b)$
$10a + 8b$	$2(5b - 4a)$
$10a - 8b$	$2(5a + 4b)$

2   Factorise:

a $2a + 4$

b $3b + 15$

c $5c - 10$

d $7d - 21$

e $4e + 6$

f $12 - 2f$

g $g^2 + 6g$

h $8j - 2k$

i $2x + 4y - 6z$

j $22m - 11n - 33p$

k $pq + pr - pt$

l $x^2 - 7x$

m $y^2 + y$

n $3z^2 + 4z$

o $2a - 3a^2$

p $4c^2 - 6c$

q $2d^2 + 8d$

r $5e + 20e^2$

s $6y^2 - 3y + 9xy$

AQA *Examiner's tip*

It helps to think of y^2 as $y \times y$ when you are factorising.

3   Explain why the expression $5x + 11y$ cannot be factorised.

4   Lisa says that $7p + 2q$ can be factorised as $2(3\frac{1}{2}p + q)$.

Is she correct?

Explain your answer.

5   Factorise each set of expressions to find the 'odd one out'.

a $3x + 12$ $2x - 8$ $4x + x^2$

b $3 - 3y^2$ $2y - 2y^2$ $5 - 5y$

c $6p^2 + 4q^2$ $6p + 4q$ $3p^2 + 2pq$

6 n is an integer.

Decide whether each of the following statements is true or false.

a $3n + 15$ is always a multiple of 3

b $4n - 1$ is never a multiple of 4

c $5n - 1$ is never prime

F

5 Assess (k!)

1 Simplify:

a $x + x + x$ g $7e - e - 2e + 4e$

b $y + y - y - y + y$ h $2f + 2g + 7f - 3g$

c $4a + 6a - 5a$ i $10h - 2j - 5h - 4j$

d $3b - b + 9b - 4b$ j $4k + 7m - 5m - k$

e $c + 12c - 3c - 4c$ k $n - 2p + 3n - 2p$

f $5d + 11d - 2d - 4d$ l $8q + r + 4r - 7q$

2 Cara has to simplify $7x - 3y - y - 2x$
She writes down $5x + 4y$ as her answer.
Describe her mistake and find the correct answer.

3 $p = 10$ and $q = 7$
Find the value of:

a $3p + 2q$ d $q - p$

b $2q + 3p$ e pq

c $p - q$ f $5pq$

4 $r = 9$ and $t = 0$
Find the value of:

a $r + t$ d $t - r$

b $2r + 3t$ e rt

c $5r - 7t$ f $7rt$

E

5 $u = -6$ and $v = -5$
Find the value of:

a $u + v$ d $2u - 3v$

b $u - v$ e $5u - 6v$

c $v - u$ f $\dfrac{5u}{v}$

6 Greg says that $3x + 2y$ will always have the same value as $2x + 3y$
Give an example to show that Greg is wrong.

7 a   When $m = 2$ and $n = 5$, $5n - m = 23$
     Write down two different expressions in m and n that have the value 23.

    b   Ellie writes down $nm - 2$ as one of her answers to part **a**.
     Can you spot Ellie's mistake?

D

8 Multiply out:

a $7(a - 2)$ d $5(e - 3f)$ g $x(x + 2)$

b $3(2b - 1)$ e $2m(n + 2)$ h $y(3 - y)$

c $4(c + d)$ f $3t(2u - 3)$ i $5z(z - 1)$

9 Factorise:

a $3a - 12$

b $2b + 10$

c $9 - 6c$

d $14 - 7d$

e $4p - 2q + 8r$

f $12x + 6y - 4z$

g $x^2 - 4x$

h $yz + z^2$

i $w^2 + vw - 3w$

10 Expand and simplify:

a $3(a + 2) + 2(a - 1)$

b $3(b - 2) + 3(b + 5)$

c $2(3c + 4) - 3(c - 4)$

d $5(x + 1) + 6(x + 7)$

e $6(2y - 3) - 4(3y - 2)$

f $8(w - 1) + 2(4 + 2w)$

g $2(x + y) + 5(2x + y)$

h $3(4p - q) + 2(p - 3q)$

i $5(2a + 3b) - 2(3a + 4b)$

j $4(m - 3n) - (3m - 5n)$

11 Show that $5(x + 1) + 2(5 - x) = 3(x + 5)$

AQA Examination-style questions

1
a Simplify $7a - 5b + 2 + 4a - 3b - 11$ (*3 marks*)

b Given that $f = 8$, $g = -3$ and $h = \frac{1}{2}$, work out the value of $4f - 5g - 2h$ (*3 marks*)

AQA 2008

6 Coordinates

Objectives

Examiners would normally expect students who get these grades to be able to:

G

use coordinates in the first quadrant

F

use coordinates in all four quadrants

E

draw lines such as $x = 3$ and $y = x$

C

find the coordinates of the midpoint of a line segment.

Key terms

coordinates
axis (pl. axes)
origin
horizontal axis
vertical axis
quadrant
line segment
midpoint

Did you know?

Cartesian coordinates

Cartesian coordinates were invented by the French mathematician, René Descartes. Descartes was born in France in 1596 and was known as the 'The Father of Modern Mathematics'.

One night, as he was lying in bed, he noticed a fly on the ceiling. He wondered how he might describe the position of the fly on the ceiling and invented Cartesian coordinates.

You should already know:

✓ negative numbers

✓ number lines

Learn... 6.1 Coordinates in four quadrants

In two dimensions, each point has two **coordinates**.

The first is the *x*-coordinate.
The second is the *y*-coordinate.

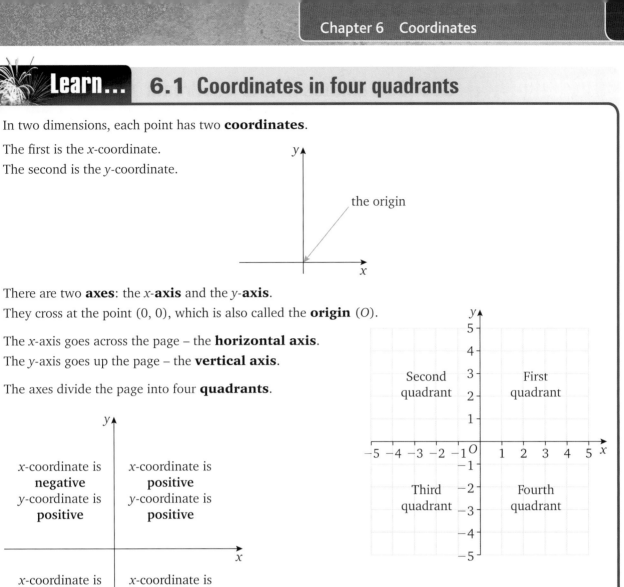

There are two **axes**: the *x*-**axis** and the *y*-**axis**.
They cross at the point (0, 0), which is also called the **origin** (*O*).

The *x*-axis goes across the page – the **horizontal axis**.
The *y*-axis goes up the page – the **vertical axis**.

The axes divide the page into four **quadrants**.

x-coordinate is **negative** *y*-coordinate is **positive**	*x*-coordinate is **positive** *y*-coordinate is **positive**
x-coordinate is **negative** *y*-coordinate is **negative**	*x*-coordinate is **positive** *y*-coordinate is **negative**

Example: Give the coordinates of the point drawn on this diagram.

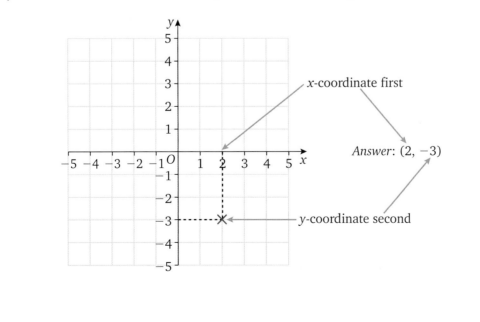

x-coordinate first

Answer: (2, −3)

y-coordinate second

Practise... **6.1 Coordinates in four quadrants** *k!* G F E D C

G
F

1 Write down the coordinates of points A, B, C, D and E.

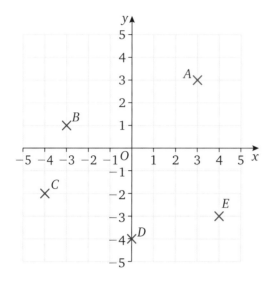

2 Draw a grid like the one in Question 1, with the x-axis and y-axis labelled from −5 to 5.

a On your grid, mark the points A(2, 4), B(4, −1), C(−1, −3) and D(−3, 2).

b Join A to B, B to C, C to D and D back to A.

c What is the mathematical name of the shape you have drawn?

F

3 Paul says that the two points E and F marked on the grid below are (−2, −2) and (−4, −2). Is he right? Give a reason for your answer.

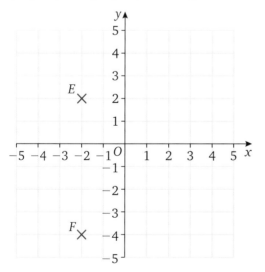

4 Write down the coordinates of:

a the top left-hand corner of the hexagon

b the bottom right-hand corner of the hexagon

c the centre of the hexagon.

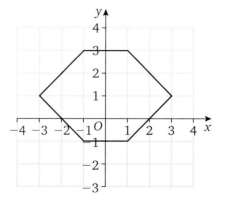

5 Judi says it is further from $(-1, 3)$ to $(5, 7)$ than it is from $(0, -1)$ to $(3, 5)$.

Plot the points on a grid and measure the distances to find out whether she is correct.

6 Draw a grid with the x-axis and the y-axis labelled from -6 to 6.

a On your grid, mark the points $P(-2, -2)$, $Q(-1, 2)$ and $R(5, 2)$.

b Join P to Q and join Q to R.

c $PQRS$ is a parallelogram.
Find the coordinates of S.

7 **a** $A(2, 0)$ and $B(2, 4)$ are two corners of a square, $ABCD$.
Write down the coordinates of C and D.
(There are two possible answers to this question.)

> **AQA** *Examiner's tip*
>
> Draw a set of axes and plot the positions of A and B.

b AB is a diagonal of the square $APBQ$.
Write down the coordinates of P and Q.

8 Andy, Ben and Chris all draw a grid and mark the points $(2, 3)$, $(4, -1)$ and $(0, 0)$.

Their teacher tells them to mark a fourth point so that they have the four corners of a parallelogram.

Andy marks $(-2, 4)$.
Ben marks $(2, -4)$.
Chris marks $(6, 2)$.

a Who is correct?

b Plot all six points.
What do you notice?

9 A yacht sends out a call for help.

Three boats, Pollyanna, Quicksilver and Roamer, are nearby and hear the call.

a Draw a grid with both axes labelled from -6 to 6.

b The yacht is at $(0, 0)$.
Mark the position of the yacht.

c Pollyanna is at $(5, 2)$.
Quicksilver is at $(-4, 4)$.
Roamer is at $(-3, 5)$.
Mark the positions of the three boats.

d Which boat is closest to the yacht?

Learn... 6.2 Introduction to straight-line graphs

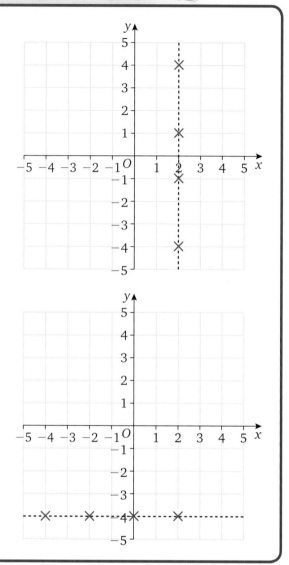

Vertical lines

The points $(2, 4)$, $(2, 1)$, $(2, -1)$, and $(2, -4)$ are marked on the grid.

Any point on the line through these points will have an x-coordinate of 2.

The equation of this line is $x = 2$

Horizontal lines

The points $(-4, -4)$, $(-2, -4)$, $(0, -4)$ and $(2, -4)$ are marked on the grid.

Any point on the line through these points will have a y-coordinate of -4.

The equation of this line is $y = -4$

6.2 Introduction to straight-line graphs

Practise...

G F E D C

E

1 Use the vertical line graph above to write down the coordinates of two different points on the line $x = 2$

2 Use the horizontal line graph above to write down the coordinates of two different points on the line $y = -4$

3 Write down the coordinates of three points on the line $x = -3$

4 Write down the coordinates of three points on the line $y = 2$

5 Write down the coordinates of three points on the line $y = -1$

6 **a** Write down the coordinates of three points on the x-axis.

   b Wayne says that the x-axis is the line $x = 0$
   Use your answer to part **a** to explain why Wayne is wrong.

   c What is the correct equation for the x-axis?

E

7 The line $x = 7$ crosses the line $y = 3$ at the point P.

Write down the coordinates of P.

AQA **Examiner's tip**

Drawing a sketch may help you to answer these questions.

8 The line $x = -5$ crosses the line $y = -4$ at the point Q.

Write down the coordinates of Q.

9 Write down the coordinates of the point where the line $y = -2$ crosses the y-axis.

10 Write down the coordinates of the point where the line $x = 4$ crosses the x-axis.

11 The line $x = -6$ does not cross the line $x = 1$. Why?

12 Write down the equations of:

a the line PQ

b the line ST

c the line UV.

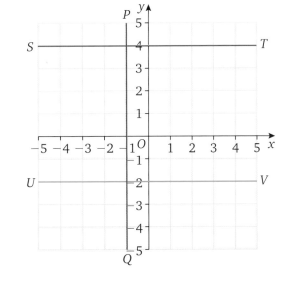

13 Draw a grid like the one in Question 12, with the axes numbered from -5 to 5.

On your grid, plot the points $A(-4, -4)$, $B(-1, -1)$ and $C(2, 2)$.

Draw the straight line through these three points.

a Write down the coordinates of two other points on this line.

b If you make the line longer, will it go through the point (3, 4)?

Give a reason for your answer.

c Jodie says the equation of the line is $x = -4$

Geeta says its equation is $x = y$

Lily says its equation is $x = -y$

Who is correct?

Give a reason for your answer.

14 Draw a coordinate grid with both axes labelled from -6 to $+6$.

a Draw a flag by joining (2, 1) to (2, 2) to (2, 4) to (4, 3) and back to (2, 2).

b i Change the signs of the x-coordinates and draw the flag again.

ii What happens to the flag?

c i Change the signs of both coordinates and draw the flag again.

ii What happens to the flag?

Learn... **6.3 The midpoint of a line segment**

A **line segment** is the part of a line joining two points.

The coordinates of the midpoint are the means of the coordinates of the end points.

Example: A line segment has been drawn from $A(-4, 1)$ to $B(2, 3)$.

Find the **midpoint** of AB.

Solution: **Method 1**

Measure halfway along the line.

The midpoint of the line is at $(-1, 2)$.

Method 2

Find the mean of the coordinates of the end points.

x: $\dfrac{-4 + 2}{2} = -1$

y: $\dfrac{1 + 3}{2} = 2$

Add the two x-coordinates and divide by 2.

Add the two y-coordinates and divide by 2.

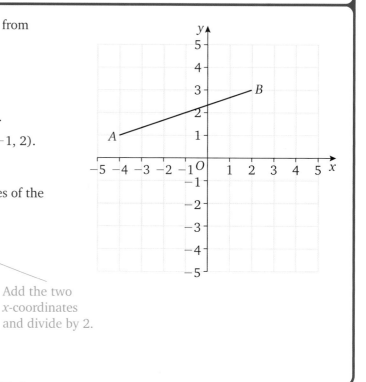

6.3 The midpoint of a line segment

Practise...

C

1 **a** Work out the coordinates of the point halfway between $(2, 5)$ and $(-4, 1)$.

b Draw a grid with the x-axis and the y-axis labelled from -5 to 5.

Plot the points $(2, 5)$ and $(-4, 1)$.

Use your diagram to check your answer to part **a**.

> **Bump up your grade**
>
> To get a Grade C you need to be able to find the coordinates of the midpoint of a line segment.

2 Work out the coordinates of the point halfway between $(0, 4)$ and $(2, 6)$.

3 A is the point $(3, -1)$ and B is the point $(-5, -5)$.

Work out the coordinates of the midpoint of the line AB.

4 Lincoln says that the point $(1, 2\frac{1}{2})$ is halfway between $(-4, 3)$ and $(6, -8)$.

Is he correct?

Give a reason for your answer.

> **AQA** **Examiner's tip**
>
> It often helps to sketch a diagram and put the points on it. This also gives you a quick check on your calculations.

5 R is the midpoint of the line PQ.

The coordinates of Q are $(3, 2)$.

R is the point $(1, 1)$.

What are the coordinates of P?

6 $A(2, 5)$, $B(5, -2)$ and $C(-2, 2)$ are the vertices of a triangle.

   a   Find the coordinates of M, the midpoint of AB.

   b   Find the coordinates of N, the midpoint of BC.

   c   Draw a grid with the x-axis and y-axis labelled from -3 to 6.
      Plot the points A, B, C, M and N.

   d   Draw the lines MN and AC.
      What do you notice about them?

7   A quadrilateral $PQRS$ has these coordinates:
   $P(0, 4)$; $Q(6, 2)$; $R(1, -3)$; $S(-5, -1)$.

   a   Find the midpoint of the diagonal PR.

   b   Find the midpoint of the diagonal QS.

   c   What do your results tell you about the quadrilateral $PQRS$?

8   The quadrilateral $TUVW$ is a kite.
   Plot $T(1, 3)$, $U(3, 3)$ and $W(-4, -4)$ on a grid.
   Find the coordinates of the fourth vertex, V.

6 Assess 🔊

1   Draw a grid with x- and y-axes labelled from 0 to 5.

   a   Plot the points $A(1, 4)$, $B(4, 4)$, $C(3, 2)$ and $D(0, 2)$.

   b   Join A to B, B to C, C to D and D to A.

   c   Which line is the same length as AB?

   d   Which line is parallel to AB?

2   Draw a grid with the x-axis labelled from -4 to 2 and the y-axis from -3 to 5.

   a   Plot the points $P(-1, 4)$, $Q(1, -2)$ and $R(-3, -2)$.

   b   Join PQ, QR and RP to form a triangle.

   c   Which two sides of your triangle are equal?

3 Ten points are marked on the grid below.

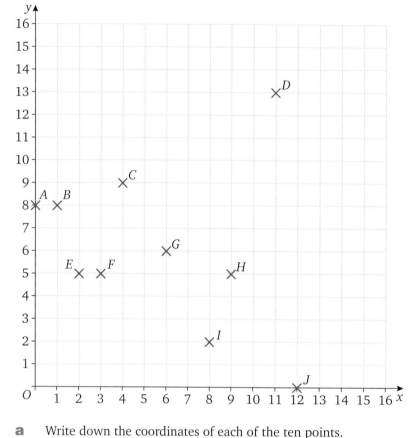

a Write down the coordinates of each of the ten points.

b Which point lies on the *x*-axis?

c Which point is furthest from the origin?

d Sean says that the points *E*, *F* and *H* lie on the line *x* = 5.
What mistake has Sean made?

e Which point lies on the line *y* = *x*?

4 The line *x* = 4 crosses the line *y* = −1 at the point *T*.

Write down the coordinates of *T*.

5 **a** Write down the coordinates of three points that lie on the *y*-axis.

b Write down the coordinates of three points that lie on the line *y* = −*x*

6 Draw a grid with the *x*-axis and the *y*-axis labelled from −4 to 5.

Plot the points *P*(−2, 2), *Q*(4, 4) and *R*(4, −1).

Join *P* to *Q* and *Q* to *R*.

a *PQRS* is a parallelogram.
S is a point in the third quadrant.
Find the position of *S* and write down its coordinates.

b Use your diagram to write down the coordinates of the midpoint of *PQ*.

7 Draw a grid with the *x*-axis and the *y*-axis labelled from −6 to 6.

Plot the points *A*(5, 2), *B*(2, −5) and *C*(−2, −1).

Join *AB*, *BC* and *AC* to form a triangle.

Find the midpoint of *BC* and label it *M*.

Write down the coordinates of *M*.

Join *AM*.

What is the angle between *AM* and *BC*?

8 *D* is the point (1, 4) and *E* is the point (3, −3).

What are the coordinates of the midpoint of *DE*?

9 *W* is the midpoint of the line *UV*.

U is the point (−3, 0).

W is the point (1, 2).

What are the coordinates of *V*?

AQA Examination-style questions

1 A shape *ABCD* is drawn on the grid.

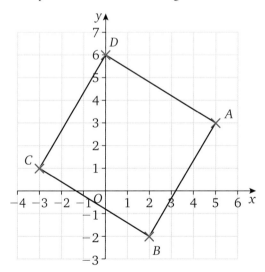

a Write down the coordinates of *A*.   (*1 mark*)

b Write down the coordinates of *C*.   (*1 mark*)

c i Draw the line *AC*. Mark the midpoint of *AC* and label it *M*.   (*1 mark*)
ii Write down the coordinates of *M*.   (*1 mark*)

d Explain why *M* is the same distance from *B* as it is from *D*.   (*1 mark*)

AQA 2008

Objectives

Examiners would normally expect students who get these grades to be able to:

F

solve a simple equation such as $5x = 10$ or $x + 4 = 7$

E

solve an equation involving fractions such as $\frac{x}{3} = 4$ or $2x - 3 = 8$

D

solve more complicated equations such as $3x + 2 = 6 - x$ or $4(2x - 1) = 20$

represent and interpret inequalities on a number line

C

solve an equation such as $4x + 5 = 3(x + 4)$ or $\frac{x}{2} - \frac{x}{8} = 9$ or $\frac{2x - 7}{4} = 1$

solve an inequality such as $2x - 7 < 9$

find the integer solutions of an inequality such as $-8 < 2n \leqslant 5$

How do we start this thing?

Did you know?

'Ink blots to space rockets'

$\bigcirc + 3 = 21$

What is the number under the blob?

You can guess the answer without knowing any algebra.

You can't design a space rocket by guesswork, but this chapter will show you how to take the first steps in solving complicated equations.

Then you might end up designing the next space rocket.

Key terms

unknown	inequality
solve/solution	$<$ (less than)
operation	$\leqslant$ (less than or equal to)
inverse operation	$>$ (greater than)
brackets	$\geqslant$ (greater than or equal to)
denominator	integer

You should already know:

✔ how to collect like terms

✔ how to use substitution

✔ how to multiply out brackets by a positive or negative number.

Learn... 7.1 Simple equations

Equations are used when you are trying to find an **unknown** value. Follow these steps to find the value of the unknown. This is called the **solution**.

- Think about the **operations** $(+, -, \times, \div)$ that have been applied to x.

- Reverse these operations, doing the same to both sides of the equation.

- Where there are two operations, reverse the second operation first, e.g. $2x + 3$ means 'multiply x by 2, then add 3', so the **inverse operations** will be 'subtract 3, then divide by 2'.

Example: Solve the equation:

$3x = 21$ Remember that $3x$ means $3 \times x$

Solution: $\dfrac{3x}{3} = \dfrac{21}{3}$ Divide both sides by 3.

$x = 7$

AQA *Examiner's tip*

Check your answer by substituting it back into the equation to see whether it fits, e.g. 3×7 does equal 21.

Example: Solve the equation:

$x + 9 = 4$

Solution: $x + 9 - 9 = 4 - 9$ Subtract 9 from both sides.

$x = -5$

Example: Solve the equation:

$\dfrac{x}{8} = 1$

Solution: $\dfrac{x}{8} \times 8 = 1 \times 8$ Multiply both sides by 8.

$x = 8$

Example: Solve the equation

$5x - 2 = 13$

Solution: This is an equation with two operations, $\times$ and $-$.

Reverse the 'subtract 2' operation first:

$5x - 2 + 2 = 13 + 2$ Add 2 to both sides.

$5x = 15$

$\dfrac{5x}{5} = \dfrac{15}{5}$ Divide both sides by 5.

$x = 3$

Practise... 7.1 Simple equations

G F E D C

1 Solve these equations.

a	$2x = 12$	**d**	$4a = 6$	**g**	$x - 3 = 8$	**j**	$a + 9 = 2$
b	$3y = 18$	**e**	$6b = 15$	**h**	$y + 4 = 12$	**k**	$b - 1 = -3$
c	$5z = 35$	**f**	$2c = -4$	**i**	$z + 1.5 = 4.8$	**l**	$c + 7.9 = 2.2$

F

E

2   Solve these equations.

a $\dfrac{x}{2} = 6$ i $23 = 4z + 7$

b $\dfrac{y}{4} = 5$ j $3 = 2a + 9$

c $\dfrac{z}{5} = 1$ k $6 = 5b + 11$

d $\dfrac{a}{10} = 0.4$ l $8c - 15 = 9$

e $\dfrac{b}{7} = 0$ m $15 - 2x = 9$

f $\dfrac{c}{3} = -2$ n $7 - 3y = 10$

g $3x - 5 = 13$ o $4 = 13 - 6z$

h $2y + 1 = 9$ p $0 = 28 - 4t$

> **Hint**
>
> If the unknown is on the right side of the equation, swap it around before you start to solve it.

3   a   Make up five different equations that have the answer $z = 3$

Use a different style for each equation:

- one which requires division
- one which requires multiplication
- one which requires addition
- one which requires subtraction
- one which requires a combination of operations.

b   Make up three different equations that have the answer $t = -5$

4   Tony thinks of a number, doubles it and adds 11.
The answer is 19.
Write this as an equation.
Solve the equation to find Tony's number.

5   Jackie thinks of a number, multiplies it by 7 and adds 5.
The answer is 47.
Write this as an equation.
Solve the equation to find Jackie's number.

6   Sol goes out with a £5 note in his pocket.
He buys x snack bars at 40p each.
He has 20p left.
Find the value of x.

7   The angles in a triangle add up to 180°.
Write down an equation in x.

Solve your equation to find the value of x.

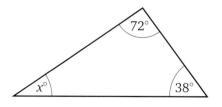

8   The angles in a quadrilateral add up to 360°.
Write down an equation in y.

Solve your equation to find the value of y.

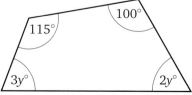

Learn... 7.2 Harder equations k!

These include equations where x appears on both sides.

Follow these steps to **solve** the equation:

- Collect together on one side all the terms that contain the **unknown** letter (x).
- Collect together on the other side all the other terms.
- Remember signs belong with the term **after** them.
- Take one step at a time – do not try to do two steps at once.

Example: Solve the equations:

a $2x + 3 = 18 - x$

b $3y + 9 = 5y - 8$

Solution:

a $2x + 3 + x = 18 - x + x$ Add x to both sides (this collects all the x terms together on the left-hand side).

$3x + 3 = 18$

$3x + 3 - 3 = 18 - 3$ Take 3 from both sides (this collects all the numbers on the right-hand side).

$3x = 15$

$\dfrac{3x}{3} = \dfrac{15}{3}$ Divide both sides by 3.

$x = 5$

b $3y + 9 - 3y = 5y - 8 - 3y$ Take $3y$ from both sides (this collects all the y terms on the right-hand side).

$9 = 2y - 8$

$9 + 8 = 2y - 8 + 8$ Add 8 to both sides.

$17 = 2y$

$\dfrac{17}{2} = \dfrac{2y}{2}$ Divide both sides by 2.

$8.5 = y$

$y = 8.5$ Write the equation with y on the left.

AQA *Examiner's tip*

Collect the terms in y on the side that has the largest number of them already.

Practise... 7.2 Harder equations k!

G F E D C

1 Solve these equations.

a $4x + 1 = 2x + 13$ e $6p + 2 = 9 + 4p$ i $7c - 1 = 3 - c$

b $2y - 3 = y + 4$ f $3 + q = 17 - 6q$ j $25 + 2d = 5d + 4$

c $5z - 2 = 8 + 3z$ g $7 + 2a = 2 - 3a$ k $6 - 7e = 3 - 6e$

d $t + 3 = 9 - 3t$ h $8b - 3 = 2b - 15$ l $5f + 10 = 2 + f$

D

2 Jared solves the equation $9x - 2 = 5 - 4x$

His first step is $5x - 2 = 5$

What mistake has Jared made?

D

3 Ella solves the equation $5y + 6 = 2 - y$

She writes down $6y = 4$

What mistake has Ella made?

4 Dean solves the equation $3x - 11 = 4 + 2x$

He gets the answer $x = 7$

Can you find Dean's mistake?

5 Rick solves the equation $2y + 5 = 3 - 3y$

He gets the answer $y = -2$

Can you find Rick's mistake?

⚠ 6 $4z - 3 = \bullet - 2z$

The answer to this equation is $z = 5$

What is the number under the blob?

⚠ 7 $2a + \bullet = 5 - 7a$

The answer to this equation is $a = -1$

What is the number under the blob?

⚠ 8 **a** If $b = 11$, find the value of $3b - 8$

b Using your answer to part **a**, explain why $b = 11$ is **not** the solution of the equation $3b - 8 = 19 - 2b$

⚠ 9 **a** If $c = -4$, find the value of $9 - 5c$

b Using your answer to part **a**, explain why $c = -4$ is **not** the solution of the equation $6c + 13 = 9 - 5c$

Learn... 7.3 Equations with brackets

For equations such as $3(2x - 1) = 12$ your first step is to deal with the **bracket**.

This usually means multiplying out the bracket.

In the first example below, you could start with division instead.

Example: Solve the equations:

a $4(3x - 1) = 32$

b $7 - 3(y + 2) = 5 - 4y$

Solution: **a** $4(3x - 1) = 32$

Multiply out the bracket first, then follow the rules for solving equations.

$12x - 4 = 32$ Remember to multiply **both** terms in the bracket by 4.

$12x - 4 + 4 = 32 + 4$ Add 4 to both sides.

$12x = 36$

$\dfrac{12x}{12} = \dfrac{36}{12}$ Divide both sides by 12.

$x = 3$

Alternative method:

$$4(3x - 1) = 32$$

$3x - 1 = 8$	Divide both sides by 4.
$3x - 1 + 1 = 8 + 1$	Add 1 to both sides.

$$3x = 9$$

$\dfrac{3x}{3} = \dfrac{9}{3}$	Divide both sides by 3.

$$x = 3$$

This alternative method works because 4 is a factor of 32.
It cannot be used for all equations with brackets, as the next example shows.

b $7 - 3(y + 2) = 5 - 4y$

Multiply out the bracket first, then follow the rules for solving equations.

$7 - 3y - 6 = 5 - 4y$	Note: $-3 \times +2 = -6$
$1 - 3y = 5 - 4y$	The numbers on the left-hand side have been collected.
$1 - 3y - 1 = 5 - 4y - 1$	Subtract 1 from both sides.
$-3y = 4 - 4y$	
$-3y + 4y = 4 - 4y + 4y$	Add 4y to both sides.
$y = 4$	

> **AQA Examiner's tip**
>
> Don't try to do two steps at once – most students make mistakes if they rush their working.

Practise... 7.3 Equations with brackets 🔁 G F E D C

1 Solve these equations.

a $5(x + 3) = 55$ **d** $4(a + 1) = 24$

b $2(y - 4) = 16$ **e** $7(b - 2) = 7$

c $9 = 3(z - 7)$ **f** $13 = 2(c + 5)$

> **Bump up your grade**
>
> You need to be able to solve equations which have brackets **and** the unknown occurring twice to get a Grade C.

2 Solve these equations.

a $4(p + 2) = 2p + 9$ **f** $6 + c = 5(c - 2)$ **k** $3(y - 4) + 2(4y - 2) = 6$

b $6(q - 3) = 17 - q$ **g** $11d - 1 = 3(d + 1)$ **l** $10 - 3(k + 2) = 7 - k$

c $2(5t - 1) = 13$ **h** $2(1 - 2e) = 5 - 3e$ **m** $23 = 6 - 4(t - 5)$

d $5a + 3 = 4(a - 2)$ **i** $2 - 5f = 3(2 - f)$ **n** $4(p - 3) - 3(p - 4) = 14$

e $3(2b - 3) = 1 + 7b$ **j** $6(2 + 3x) = 11x + 5$ **o** $2(q - 9) - (7q - 3) + 25 = 0$

3 Natalie thinks of a number, adds 7 and then doubles the result.
Her answer is 38.
Write this as an equation.
Solve the equation to find Natalie's number.

4 Rob thinks of a number, subtracts 5 and then multiplies the result by 4.
His answer is 32.
Write this as an equation.
Solve the equation to find Rob's number.

Learn... 7.4 Equations with fractions

At some stage in solving an equation with a fraction, you have to clear the fraction by multiplying both sides by the **denominator**.

For example, if the equation contains $\frac{x}{3}$, you will multiply by 3.

If there is more than one fraction, say $\frac{3x}{5}$ and $\frac{x}{2}$, you will multiply by both denominators.

In this case, this is $5 \times 2 = 10$

Harder equations have more than one term on the top of the fraction. There is an 'invisible bracket' around the terms on top of an algebraic fraction.

Example: Solve the equation

$$\frac{x}{3} - 2 = 5$$

Solution: This is an example of the simplest type of equation with a fraction.

$$\frac{x}{3} - 2 + 2 = 5 + 2 \qquad \text{Start by adding 2 to both sides.}$$

$$\frac{x}{3} = 7 \qquad \text{Now the fraction term is on its own.}$$

$$\frac{x}{3} \times 3 = 7 \times 3 \qquad \text{Multiply both sides by 3 (the \textbf{denominator}).}$$

$$x = 21$$

Example: Solve the equation

$$\frac{3x}{5} - \frac{x}{2} = 1$$

Solution: This is an example where there is more than one fraction.

You need to multiply by both denominators, in this case, $5 \times 2 = 10$

Multiply each term by 10.

$$10 \times \frac{3x}{5} = \frac{30x}{5} = 6x$$

$$10 \times \frac{x}{2} = \frac{10x}{2} = 5x$$

$$10 \times 1 = 10$$

$$6x - 5x = 10$$

$$x = 10$$

> **AQA Examiner's tip**
>
> Don't forget to multiply the right-hand side as well as the left-hand side.

Example: Solve the equation

$$\frac{5x + 2}{4} = 3$$

Solution: This is an example of a harder equation with more than one term on the top of the fraction.

$\dfrac{(5x + 2)}{4}$ is the same as one-quarter of $(5x + 2)$

Multiply by 4 to get $5x + 2$

$$5x + 2 = 12 \longleftarrow 4 \times 3$$

$$5x + 2 - 2 = 12 - 2 \qquad \text{Subtract 2 from both sides.}$$

$$5x = 10$$

$$\frac{5x}{5} = \frac{10}{5} \qquad \text{Divide both sides by 5.}$$

$$x = 2$$

> **Hint**
>
> You should put in the invisible bracket before you start your working.

Practise...   **7.4 Equations with fractions** 🔑   G F E D C

C

1   Solve these equations.

a $\dfrac{x}{2} - 5 = 4$ **f** $\dfrac{c}{6} + 5 = 2$ **k** $\dfrac{3q + 8}{2} = 13$

b $\dfrac{y}{5} + 3 = 7$ **g** $\dfrac{4x + 1}{3} = 11$ **l** $\dfrac{4t - 3}{3} = 7$

c $5 = 1 + \dfrac{z}{3}$ **h** $\dfrac{2y - 7}{5} = 3$ **m** $\dfrac{x}{5} + \dfrac{x}{3} = 8$

d $7 + \dfrac{a}{3} = 8$ **i** $1 = \dfrac{9 - z}{3}$ **n** $\dfrac{y}{2} - \dfrac{y}{8} = 3$

e $9 - \dfrac{b}{2} = 2$ **j** $\dfrac{p + 3}{4} = 5$ **o** $\dfrac{5z}{6} - \dfrac{7z}{12} = 4$

2   **a** $\dfrac{5a - 1}{2} = a - 5$ **c** $c - 7 = \dfrac{11 - c}{3}$ **e** $\dfrac{q}{3} - \dfrac{1}{4} = \dfrac{q}{6}$

    b $\dfrac{2b - 5}{8} = 5 - b$ **d** $\dfrac{3p}{2} = 5 - \dfrac{p}{6}$ **f** $\dfrac{3t}{8} + \dfrac{1}{4} = \dfrac{2t}{5}$

3   Faria says the answer to the equation $\dfrac{x + 2}{5} = 4 - x$ is $x = 9$

Use substitution to check whether Faria is correct.

4   Ed and Gary solve the equation $\dfrac{4y - 3}{5} = 2y + 3$

Ed gets the answer $y = -2$ and Gary gets $y = -3$

Check their answers to see which of them is correct.

5   The equation $\dfrac{6p - 5}{2} = 4 + 3p$ cannot be solved. Why?

Learn...   **7.5 Inequalities and the number line**

The four **inequality** symbols are:

<	⩽	>	⩾
less than	less than or equal to	greater than	greater than or equal to

A number line shows the range of values for x.

An open circle shows that the range does not include that end of the line.

For example $x > 1$ or $y < 5$

A closed circle shows that the range does include that end of the line.

For example $x \leqslant 3$ or $y \geqslant 5$

This is the number line for $x > 1$

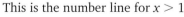

x could be any number greater than 1... *but not 1.*

The open circle shows that x can be close to 1 but not equal to 1.

This is the number line for $x \leqslant 3$

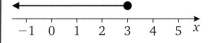

x could be any number less than or equal to 3.

The closed circle shows that x can be equal to 3.

This is the number line for $x < -1$ or $x \geqslant 2$

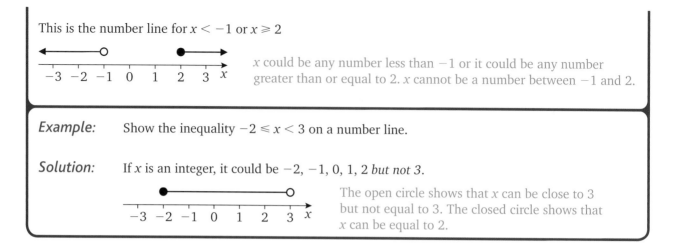

x could be any number less than -1 or it could be any number greater than or equal to 2. x cannot be a number between -1 and 2.

Example: Show the inequality $-2 \leqslant x < 3$ on a number line.

Solution: If x is an integer, it could be $-2, -1, 0, 1, 2$ *but not 3.*

The open circle shows that x can be close to 3 but not equal to 3. The closed circle shows that x can be equal to 2.

Practise...   **7.5 Inequalities and the number line**

D

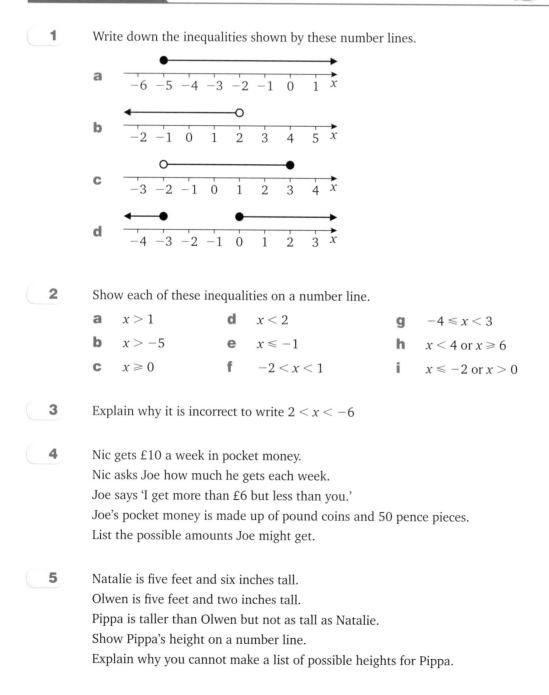

1   Write down the inequalities shown by these number lines.

2   Show each of these inequalities on a number line.

a $x > 1$ d $x < 2$ g $-4 \leqslant x < 3$

b $x > -5$ e $x \leqslant -1$ h $x < 4$ or $x \geqslant 6$

c $x \geqslant 0$ f $-2 < x < 1$ i $x \leqslant -2$ or $x > 0$

3   Explain why it is incorrect to write $2 < x < -6$

4   Nic gets £10 a week in pocket money.
Nic asks Joe how much he gets each week.
Joe says 'I get more than £6 but less than you.'
Joe's pocket money is made up of pound coins and 50 pence pieces.
List the possible amounts Joe might get.

5   Natalie is five feet and six inches tall.
Olwen is five feet and two inches tall.
Pippa is taller than Olwen but not as tall as Natalie.
Show Pippa's height on a number line.
Explain why you cannot make a list of possible heights for Pippa.

Learn... 7.6 Solving inequalities

Some inequalities are very similar to equations.

The inequality $3x - 2 > 4$ is similar to the equation $3x - 2 = 4$

To solve this inequality, use inverse operations as you would with the equation.

$3x - 2 + 2 > 4 + 2$ Add 2 to both sides.

$3x > 6$ Divide both sides by 3.

$x > 2$

You may be asked to list **integer** values (whole numbers) that satisfy an inequality.

For example, the integers that satisfy $-3 \leqslant x < 5$ are $-3, -2, -1, 0, 1, 2, 3, 4$.

Sometimes you have to combine these two skills, as in the example below.

Example: List all the integer values of n such that $-5 < 2n \leqslant 6$

Solution: Divide every term in the inequality by 2
$-2.5 < n \leqslant 3$

Integer values for n are: $-2, -1, 0, 1, 2, 3$

Practise... 7.6 Solving inequalities   **G F E D C**

1   Solve these inequalities.

a $3x - 2 \geqslant 4$

b $2y + 7 \leqslant 16$

c $4z + 12 < 0$

d $5 + 2p > 1$

e $8 < 2 + 3q$

f $5 > 13 - x$

2   List all the integer values of n such that:

a $-3 < n < 4$

b $1 \leqslant n < 6$

c $-5 < n \leqslant -1$

d $-2 \leqslant n \leqslant 1$

e $0 < 3n < 15$

f $-4 < 2n \leqslant 6$

g $-10 \leqslant 4n < 12$

h $-5 \leqslant 5n \leqslant 8$

3   Find the largest integer that satisfies the inequality $5 - 2x \geqslant 1$

4   Find the smallest integer that satisfies the inequality $7 < 3(2x + 9)$

5   List all the pairs of positive integers, x and y, such that $3x + 4y \leqslant 15$

6   Jiffa is 14 years old.

She says to her Uncle Asif, 'How old are you?'

He says, 'In 9 years' time I shall be more than twice as old as I was when you were born.'

Write down an inequality and solve it to find the greatest age Asif could be.

Assess   (k!)

F

1   Solve these equations.

    a $9x = 81$ **b** $y - 4 = 12$ **c** $10z = 50$

E

2   Solve these equations.

    a $\dfrac{a}{3} = 9$ **c** $2d - 5 = 6$ **e** $8 + 3f = 5$

    b $4c + 11 = 3$ **d** $5e + 4 = 12$ **f** $p + 5 = 14 - 2p$

D

3   Solve these equations.

    a $2q - 1 = 5 - q$ **c** $4 + 3n = n - 10$ **e** $49 = 7(3t - 2)$

    b $6m - 7 = 2m + 3$ **d** $5(u + 1) = 35$ **f** $3(v - 4) = 9 + 2v$

4   Write down the inequalities shown by these number lines.

    a

$$-5 \quad -4 \quad -3 \quad -2 \quad -1 \quad 0 \quad 1 \quad 2 \quad 3 \quad x$$

    b

$$-5 \quad -4 \quad -3 \quad -2 \quad -1 \quad 0 \quad 1 \quad 2 \quad 3 \quad y$$

C

5   Solve these equations.

    a $4(w - 2) + 2(3w + 1) = 44$ **c** $\dfrac{y}{4} + 3 = 7$ **e** $\dfrac{x}{5} + \dfrac{x}{10} = 6$

    b $5(2x - 3) = 7 + 4(x - 1)$ **d** $4 - \dfrac{k}{3} = 6$

6   Solve these inequalities.

    a $6a - 7 \geqslant 5$ **b** $3b + 10 < 4$

7   Find the largest integer that satisfies the inequality $2x + 3 < 17$

8   List all the integer solutions of the inequality $-8 < 3n \leqslant 9$

AQA Examination-style questions   (k!)

1   An equilateral triangle is one where all the sides are the same length.

This triangle has lengths $(4x - 2)$ cm, $(2x + 5)$ cm and $(6x - 9)$ cm.

Find the value of x that makes this triangle equilateral.

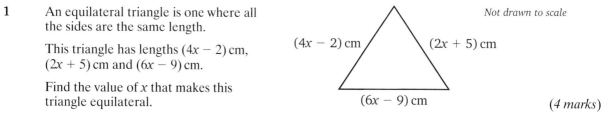

Not drawn to scale

$(4x - 2)$ cm $(2x + 5)$ cm

$(6x - 9)$ cm

(4 marks)

AQA 2009

8 Percentages

Objectives

Examiners would normally expect students who get these grades to be able to:

F

understand that percentage means 'number of parts per 100'

change a percentage to a fraction or a decimal and vice versa

E

compare percentages, fractions and decimals

work out a percentage of a given quantity

D

increase or decrease by a given percentage.
For example, find the new price of a £490 TV after a 15% reduction

express one quantity as a percentage of another

C

work out a percentage increase or decrease.

Did you know?

I gave 110%!

Some sports personalities have claimed 'I gave 110%.'

Is this possible?

When is it possible to have more than 100%?

What about these?

'I got more than 100% in the maths test!'

'The price went up by more than 100%.'

'The company made a loss of more than 100%.'

'I'm more than 100% certain that it happened.'

Key terms

percentage
equivalent fractions
VAT (Value Added Tax)
rate
discount
deposit
credit
balance
interest
amount

You should already know:

✓ about place values in decimals (for example, $0.7 = \frac{7}{10}$, $0.07 = \frac{7}{100}$)

✓ how to put decimals in order of size

✓ how to simplify fractions

✓ how to write a fraction as a decimal and vice versa.

Learn... 8.1 Percentages, fractions and decimals

1% (1 per cent) means '1 part out of 100' or 'one hundredth'.

It is equivalent to the fraction $\frac{1}{100}$ and the decimal 0.01. In money it is equivalent to '1p in the £1'.

Here is a grid of 100 squares. 40 out of 100 are shaded.

This is 40% (40 hundredths) of the grid.

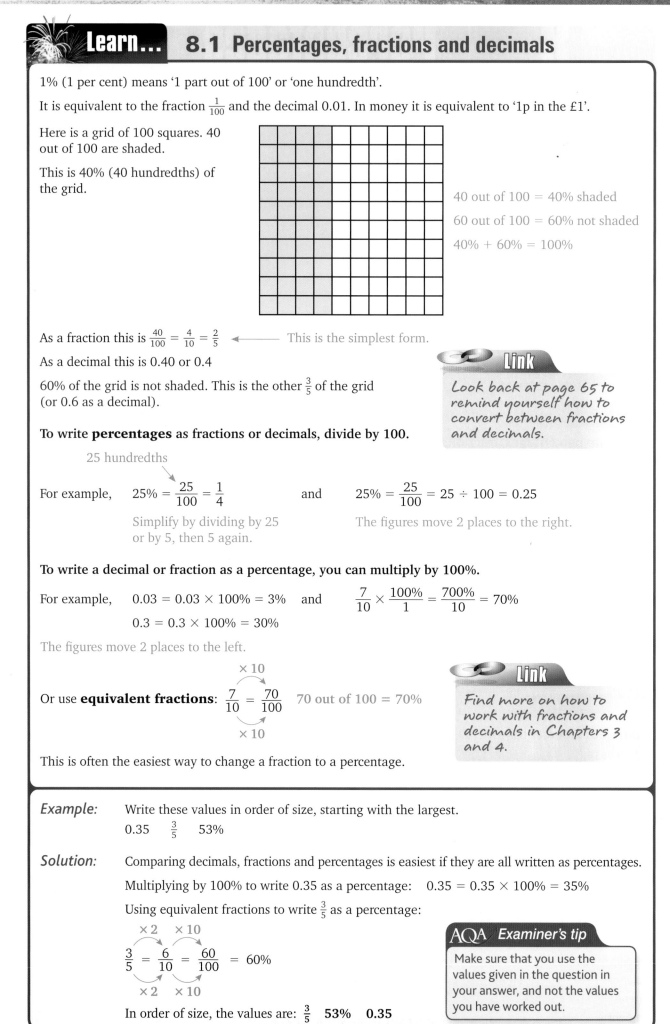

40 out of 100 = 40% shaded

60 out of 100 = 60% not shaded

40% + 60% = 100%

As a fraction this is $\frac{40}{100} = \frac{4}{10} = \frac{2}{5}$ ⟵⎯⎯⎯⎯ This is the simplest form.

As a decimal this is 0.40 or 0.4

60% of the grid is not shaded. This is the other $\frac{3}{5}$ of the grid (or 0.6 as a decimal).

To write percentages as fractions or decimals, divide by 100.

Link

Look back at page 65 to remind yourself how to convert between fractions and decimals.

25 hundredths

For example, $25\% = \dfrac{25}{100} = \dfrac{1}{4}$ and $25\% = \dfrac{25}{100} = 25 \div 100 = 0.25$

Simplify by dividing by 25 or by 5, then 5 again.

The figures move 2 places to the right.

To write a decimal or fraction as a percentage, you can multiply by 100%.

For example, $0.03 = 0.03 \times 100\% = 3\%$ and $\dfrac{7}{10} \times \dfrac{100\%}{1} = \dfrac{700\%}{10} = 70\%$

$0.3 = 0.3 \times 100\% = 30\%$

The figures move 2 places to the left.

Link

Find more on how to work with fractions and decimals in Chapters 3 and 4.

× 10

Or use **equivalent fractions**: $\dfrac{7}{10} = \dfrac{70}{100}$ 70 out of 100 = 70%

× 10

This is often the easiest way to change a fraction to a percentage.

Example: Write these values in order of size, starting with the largest.

0.35 $\frac{3}{5}$ 53%

Solution: Comparing decimals, fractions and percentages is easiest if they are all written as percentages.

Multiplying by 100% to write 0.35 as a percentage: $0.35 = 0.35 \times 100\% = 35\%$

Using equivalent fractions to write $\frac{3}{5}$ as a percentage:

× 2   × 10

$\dfrac{3}{5} = \dfrac{6}{10} = \dfrac{60}{100} = 60\%$

× 2   × 10

AQA *Examiner's tip*

Make sure that you use the values given in the question in your answer, and not the values you have worked out.

In order of size, the values are: $\frac{3}{5}$ 53%   0.35

Practise...

8.1 Percentages, fractions and decimals

G F E D C

1   For each 100 square below, write down

    i   the percentage that is shaded

    ii   the percentage shaded as a decimal

    iii the fraction that is **not** shaded.

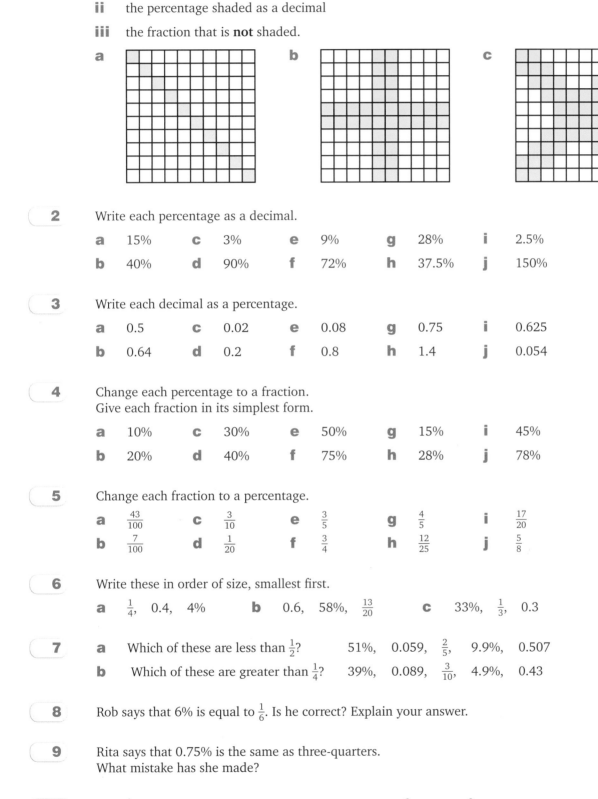

G

2   Write each percentage as a decimal.

a	15%	**c**	3%	**e**	9%	**g**	28%	**i**	2.5%
b	40%	**d**	90%	**f**	72%	**h**	37.5%	**j**	150%

G F

3   Write each decimal as a percentage.

a	0.5	**c**	0.02	**e**	0.08	**g**	0.75	**i**	0.625
b	0.64	**d**	0.2	**f**	0.8	**h**	1.4	**j**	0.054

4   Change each percentage to a fraction.
Give each fraction in its simplest form.

a	10%	**c**	30%	**e**	50%	**g**	15%	**i**	45%
b	20%	**d**	40%	**f**	75%	**h**	28%	**j**	78%

5   Change each fraction to a percentage.

a	$\frac{43}{100}$	**c**	$\frac{3}{10}$	**e**	$\frac{3}{5}$	**g**	$\frac{4}{5}$	**i**	$\frac{17}{20}$
b	$\frac{7}{100}$	**d**	$\frac{1}{20}$	**f**	$\frac{3}{4}$	**h**	$\frac{12}{25}$	**j**	$\frac{5}{8}$

E

6   Write these in order of size, smallest first.

    a $\frac{1}{4}$, 0.4, 4%   **b** 0.6, 58%, $\frac{13}{20}$ **c** 33%, $\frac{1}{3}$, 0.3

7   **a** Which of these are less than $\frac{1}{2}$?   51%, 0.059, $\frac{2}{5}$, 9.9%, 0.507

    b Which of these are greater than $\frac{1}{4}$?   39%, 0.089, $\frac{3}{10}$, 4.9%, 0.43

8   Rob says that 6% is equal to $\frac{1}{6}$. Is he correct? Explain your answer.

D

9   Rita says that 0.75% is the same as three-quarters.
What mistake has she made?

10   **a** Which of these is nearest in size to 0.4?   38% $\frac{3}{8}$ 43% $\frac{3}{7}$

     b Which of these is nearest in size to $\frac{3}{4}$?   0.7 72% $\frac{7}{9}$ 0.79

11   Find the percentage that lies halfway between $\frac{3}{4}$ and $\frac{4}{5}$

! 12 Write these fractions as percentages.

$\frac{1}{2}$ $\frac{2}{3}$ $\frac{3}{4}$ $\frac{4}{5}$ $\frac{5}{6}$ $\frac{6}{7}$ $\frac{7}{8}$ $\frac{9}{10}$

What do your results show?

? 13 Which of these do you think is the best offer?

Explain your answer.

| Buy one, get one free | 30% off | 3 for the price of 2 | $\frac{1}{3}$ off |

? 14 Jan, Kate and Leo have done a maths test.
Jan got 75% of the marks that Leo got. Kate got $\frac{4}{5}$ of the marks that Leo got.

Who got fewer marks, Jan or Kate? You **must** show your working.

Learn... 8.2 Finding a percentage of a quantity

In this unit you will need to find percentages without a calculator.

You may need to shade a percentage of a shape, for example 30% of this shape.

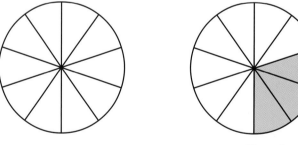

$30\% = \frac{30}{100} = \frac{3}{10}$,

so shade three out of the ten parts.

When finding a percentage of a quantity, remember that $1\% = \frac{1}{100}$ (one-hundredth)

To find 1% of something, divide it by 100.

Also remember that $10\% = \frac{10}{100} = \frac{1}{10}$ (one-tenth)

To find 10% of something, divide it by 10.

There are many other ways to find percentages of a quantity.

Without a calculator, the easiest way is usually to find 1% or 10% first.

Example: Find   **a**   40% of £320   **b**   97% of £2500

Solution: **a**   10% of £320 = £320 ÷ 10 = £32
        40% of £320 = £32 × 4 = £128   40% is 4 × 10%

AQA *Examiner's tip*

Always check whether the answer looks reasonable. As 40% is less than 50%, the answer should be less than half of £320. £128 seems reasonable.

   b   1% of £2500 = £2500 ÷ 100 = £25
        3% of £2500 = £25 × 3 = £75 ◄―――――― Here is one way to check this:
        97% of £2500 = £2500 − £75 = £2425   3% is the same as 3p in every £
                          So 3% of £2500 = 2500 × 3p
                                 = 7500p

Sometimes links with other fractions are useful.

For example,   50% = $\frac{50}{100}$ = $\frac{1}{2}$ To find 50% of a quantity, divide it by 2.

        25% = $\frac{25}{100}$ = $\frac{1}{4}$ To find 25% of a quantity, divide it by 4 (or halve it, then halve again).

Also 75% = 50% + 25%

For example, to find 75% of £420:   50% of £420 = £210   Half of £420
                     25% of £420 = £105   Half of £105
                     75% of £420 = £210 + £105 = £315   50% + 25%

Compare the above with this method: 10% of £420 = £42
                     70% of £420 = £42 × 7 = £294

5% is half of 10%, so divide £42 by 2   5% of £420 = £21

Adding 70% and 5% gives   75% of £420 = £315

Which method do you prefer? Can you think of any other ways to find 75% of £420?

Using a different method is a good way to check an answer.

8.2 Finding a percentage of a quantity

Practise...

G F E D C

1   **a**   Copy each shape onto squared paper. Shade the given percentage.
       i   25%   **ii**   80%   **iii**   10%

   b   Write down the percentage of each shape that is not shaded.

2   Find:
   a   1% of £500   **b**   4% of £500   **c**   96% of £500

Hint
96% = 100% − 4%

3   Work out:
   a   10% of £270   **e**   35% of £40   **i**   70% of 4500 litres
   b   20% of £1500   **f**   95% of £7200   **j**   5% of 80 km
   c   40% of £120   **g**   30% of 240 m   **k**   85% of 740 g
   d   15% of £840   **h**   90% of 50 kg   **l**   45% of 280 ml

G F

E

E

4　**a**　Work out:

　　　i　50% of £1800　　　**ii**　25% of £1800　　　**iii**　75% of £1800

　　b　Use a different method to check your answers to part **a**.

5　630 boys and 660 girls go to a school.
90% of the boys and 95% of the girls have a mobile phone.

　　a　How many boys have mobile phones?

　　b　How many girls have mobile phones?

　　c　How many students altogether have mobile phones?

D

6　Paula earns £840 per week.
She spends 25% of this on rent and 40% on shopping.
How much does she have left?

Hint
What % does she have left?

7　**a**　Carl explains how he finds $12\frac{1}{2}$%.
Use Carl's method to find
$12\frac{1}{2}$% of £560.

　　b　Jen says $12\frac{1}{2}$% is half of 25%.
Use Jen's method to check your
answer to part **a**.

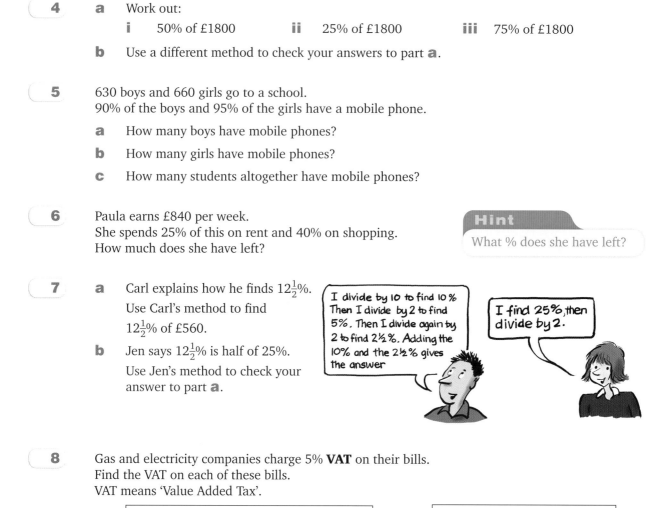

I divide by 10 to find 10%
Then I divide by 2 to find
5%. Then I divide again by
2 to find 2½%. Adding the
10% and the 2½% gives
the answer

I find 25%, then
divide by 2.

8　Gas and electricity companies charge 5% **VAT** on their bills.
Find the VAT on each of these bills.
VAT means 'Value Added Tax'.

　　a
```
Cost of electricity used
        (without VAT) = £96.40
```

　　b
```
Gas you've used
    (without VAT) = £135.80
```

9　A shop adds 17.5% VAT to all the goods it sells.
Find the VAT on each of these:

Hint
17.5% = 10% + 5% + 2.5%

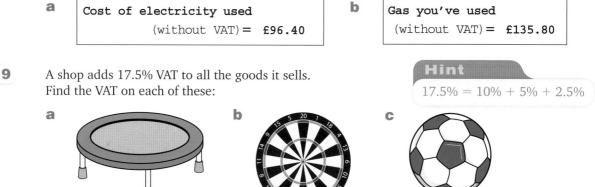

　　a　£160 plus VAT
　　b　£24 plus VAT
　　c　£16.80 plus VAT

C

10　Karen says you can find 8% of £5200 by dividing £5200 by 8.
Is this correct? Explain your answer.

11　There are 32 000 football supporters at a match.
65% of these are men, 25% are women and the rest are children.

　　a　How many more men than women are at the match?

　　b　What fraction of the football supporters at the match are children?

12　Work out:

　　a　11% of £560　　　**c**　98% of £2340　　　**e**　76% of 120 kg

　　b　29% of £84　　　**d**　42% of 750 m　　　**f**　55% of 24 litres

13   Work out:

    a $2\frac{1}{2}\%$ of 48   **b**   32.5% of 16 million   **c** $97\frac{1}{2}\%$ of £42

14   In 2009 a laptop cost £400 excluding VAT.
During 2009 the **rate** of VAT was reduced from 17.5% to 15%
How much less was the cost of the laptop including VAT after the VAT rate was reduced?

15   The table shows the tax rates in the country where Yusef lives.

Taxable income	taxed at
Up to £40 000	20%
Amount over £40 000	40%

Yusef's taxable income is £45 000.

    a   How much tax does Yusef pay?

    b   How much extra tax will Yusef pay if he gets a 2% pay rise?

16   12 800 people live in a town.
34% of these are men and 36% are women.
How many children live in the town?
You **must** show your working.

8.3 Increasing or decreasing an amount by a percentage

Learn...

It is usually best to find the increase or decrease first (using a method from Learn 8.2).

Then add or subtract the result from the original amount.

But sometimes there are quicker ways. For example, here are two ways of reducing £120 by 75%

50% of £120 = £60   50% is half of 100%   100% − 75% = 25%

25% of £120 = £30   25% is half of 50%   so the decreased amount is 25% of £120.

75% of £120 = £90   75% = 50% + 25%   50% of £120 = £60

Decreased amount = £120 − £90   25% of £120 = £30

Decreased amount = £30   Decreased amount = £30

Any method is acceptable. In this case, method 2 is quicker.
But in other examples method 1 is the best.

Example: Increase 800 by 15%

Solution: 10% of 800   = 800 ÷ 10 = 80

             5% of 800   = 80 ÷ 2   = 40   5% is half of 10%

             15% of 800   = 10% + 5% = + 80 + 40 = 120

             Increased amount = 800 + 120 = **920**

AQA *Examiner's tip*

Always check whether your answer looks reasonable. Here the answer should be a bit bigger than 800 – which it is.

Practise...

8.3 Increasing or decreasing an amount by a percentage

D

1 Increase:

a £80 by 10%   c £6000 by 5%   e £360 by 25%

b £500 by 20%   d £78 by 50%   f £1280 by 15%

2 Decrease:

a £700 by 1%   c £650 by 30%   e £250 by 90%

b £450 by 10%   d £82 000 by 3%   f £480 by 95%

3 a Increase 540 cm by 50%   c Decrease 360 km by 75%

 b Increase 280 kg by 25%   d Decrease 650 litres by 80%

Check each answer using a different method.

4 The normal price for an album is £15.
The shop reduces this by 20% in a sale.
What is the sale price?

5 A late offer gives 25% **discount** on a holiday that usually costs £720.

a What does the holiday cost after the discount?

b Check your answer using a different method.

6 Find the total cost of each of these.

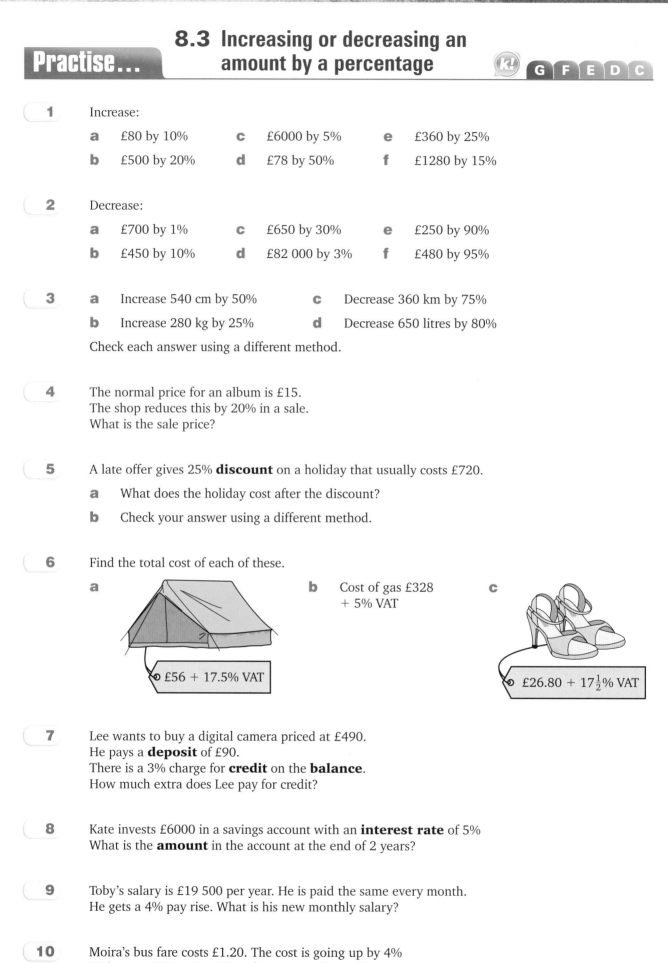

a

b Cost of gas £328
 + 5% VAT

c

£56 + 17.5% VAT

£26.80 + 17½% VAT

C

7 Lee wants to buy a digital camera priced at £490.
He pays a **deposit** of £90.
There is a 3% charge for **credit** on the **balance**.
How much extra does Lee pay for credit?

8 Kate invests £6000 in a savings account with an **interest rate** of 5%
What is the **amount** in the account at the end of 2 years?

9 Toby's salary is £19 500 per year. He is paid the same every month.
He gets a 4% pay rise. What is his new monthly salary?

10 Moira's bus fare costs £1.20. The cost is going up by 4%
Moira says 120 ÷ 4 = 30 so the new cost is £1.50.

a Explain why this is not correct.

b By what percentage has Moira increased the cost?

11 A new car costs £20 000. Its value depreciates by 10% each year.
How much will it be worth when it is 3 years old?

12 A website advertises watches at 50% off, but adds £4 for postage and packing. The final cost of buying a watch from this website is £32.
Using x to represent the original price of the watch in pounds:

  a   write down an equation for x

  b   solve the equation to find the original price of the watch.

13 A games console is advertised for sale in two shops.

Arkos	**Playshop**
£79.95 including VAT	£68 plus VAT

VAT is added at $17\frac{1}{2}\%$

Which shop is cheaper and by how much?

14 A clothes shop aims to make at least 30% profit on everything it sells.
First the manager adds 30% to the cost price of an item.
Then he increases the result to a penny less than the next pound.
So, for example, if the 30% mark-up on an item gives £27.30,
the manager prices it at £27.99.

Find the manager's price for each item in the table.

Item	Cost price
Shirt	£12
Skirt	£25
Shorts	£7
Trousers	£36
Jacket	£47

15 Last year a school had 900 girls and 880 boys.
At the beginning of this year the number of girls went down by 3%
The number of boys went up by $2\frac{1}{2}\%$

Is the total number of students greater or fewer than last year?
You **must** show your working.

Learn...

8.4 Writing one quantity as a percentage of another

To write one quantity as a percentage of another, write them as a fraction first.
Then change the fraction to a percentage. The quantities must be in the same units.

For example, to write 75 pence as a percentage of £2.50, you need to change £2.50 to 250 pence.

The percentage can then be found from $\dfrac{75}{250} = \dfrac{15}{50} = \dfrac{30}{100} = \mathbf{30\%}$

$\div 5 \quad \times 2$

You can find the percentage by changing the denominator to 100.

You can also change the fraction to a percentage by multiplying by 100%: $\dfrac{75}{250} = \dfrac{75}{250} \times \dfrac{100\%}{1} = \dfrac{7500\%}{250} = \dfrac{750\%}{25} = 30\%$

$\div 10$

This way is sometimes difficult when you don't have a calculator.

Example: 30 girls and 45 boys are members of a swimming club.

What percentage of the members are   **a**   girls   **b**   boys?

Solution: The total number of members = 30 + 45 = 75

a   30 out of 75 members are girls.

The fraction who are girls = $\frac{30}{75}$

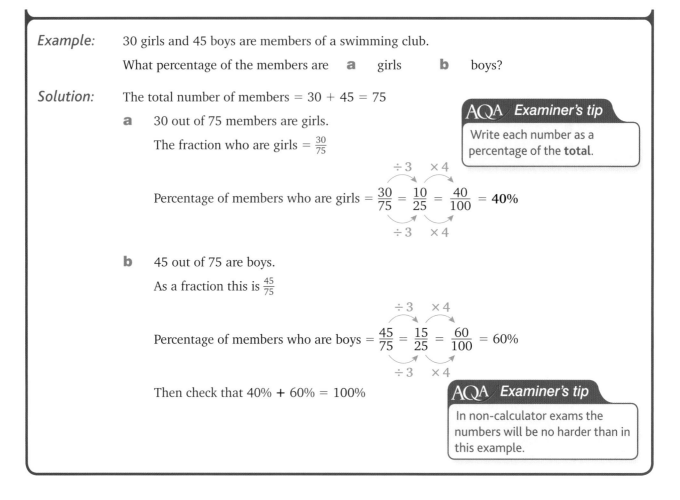

Percentage of members who are girls = $\frac{30}{75} = \frac{10}{25} = \frac{40}{100}$ = **40%**

> **AQA** *Examiner's tip*
>
> Write each number as a percentage of the **total**.

b   45 out of 75 are boys.

As a fraction this is $\frac{45}{75}$

Percentage of members who are boys = $\frac{45}{75} = \frac{15}{25} = \frac{60}{100}$ = **60%**

Then check that 40% + 60% = 100%

> **AQA** *Examiner's tip*
>
> In non-calculator exams the numbers will be no harder than in this example.

8.4 Writing one quantity as a percentage of another

Practise...

k!   G F E D C

D

1   Write:

a   £18 as a percentage of £300   **f**   £12 000 as a percentage of £60 000

b   24p as a percentage of 50p   **g**   200 g as a percentage of 250 g

c   35p as a percentage of £5   **h**   40 m as a percentage of 200 m

d   £15 as a percentage of £250   **i**   105 cm as a percentage of 140 cm

e   72p as a percentage of £1.80   **j**   1 litre as a percentage of 3 litres.

2   Fiona gets £50 for her birthday. She spends £30 and saves the rest.

a   What percentage does she spend?

b   What percentage does she save?

3   A charity has raised £1280. Their target is £4000.
What percentage of the target do they still need to raise?

4   There were 12 rainy days in June. What percentage of the month is this?

5   There are 33 girls and 27 boys in a youth club.

a   What percentage of the youth club members are   **i**   girls   **ii**   boys?

b   Show how you can check your answers to part **a**.

D

C

6 Mark gets an allowance of £25. He spends £4.25 on magazines.
What percentage of his allowance does he have left?

7 25 women and 35 men start a marathon.
18 of the women and 21 of the men complete it.

Work out:

a the percentage of women who complete the marathon

b the percentage of men who complete the marathon

c the percentage of all those taking part who do not complete it.

8 Driving lessons cost £25 each with a discount of 20% if you buy ten lessons.
So far Chris has saved up £140.
What percentage of the total cost of ten lessons has Chris saved?

9 David's maths teacher says he got 70% in his maths test.
His English teacher says he got 80 out of 120 in his English test.
In which test did David do better? Explain your answer.

10 Miss Take says '120 girls and 180 boys go to a summer camp.
What percentage of the campers are girls?'

Katya writes down $\frac{120}{180} \times \frac{100}{1}$

a What mistake has Katya made?

b What is the correct answer to Miss Take's question?

Bump up your grade

For Grade C you must be able to explain why something is wrong.

11 Write:

a 60p as a percentage of £9

b 24 cm as a percentage of 3 m

c 300 g as a percentage of 2 kg

d 200 ml as a percentage of 2 litres

e 240 cm as a percentage of 4 m

f £1800 as a percentage of £30 000

g 14 000 as a percentage of 7 million

h 8 hours as a percentage of 1 day

Hint

1 m = 100 cm
1 kg = 1000 g
1 litre = 1000 ml

! 12 What percentage of the numbers between 1 and 20 (inclusive) are:

a multiples of 6   b factors of 6   c prime numbers?

13 A gardener measures the temperature in his greenhouse every night.
The table gives the results for February.

	Mon	Tues	Wed	Thurs	Fri	Sat	Sun
Week 1	2°C	0°C	−1°C	−2°C	−4°C	−5°C	−3°C
Week 2	−2°C	0°C	2°C	4°C	5°C	4°C	1°C
Week 3	−1°C	0°C	−1°C	1°C	3°C	4°C	4°C
Week 4	−2°C	−3°C	−4°C	−3°C	−5°C	−1°C	0°C

On what percentage of the nights was the temperature

a below 0°C   b below −2°C?

14 A mobile phone costs £15 per month plus 5 pence per minute for calls.

   a Write the monthly payment as a percentage of the total cost in a month when the time on calls is:

      i 100 minutes   **iii** 300 minutes

      ii 200 minutes   **iv** 400 minutes

   b What happens to this percentage as the call time increases?

15 **a** What percentage of two-digit numbers are:

      i even numbers   **iii** multiples of 5

      ii multiples of 10   **iv** multiples of 3?

   b What percentage of two-digit numbers contains the digit 3?

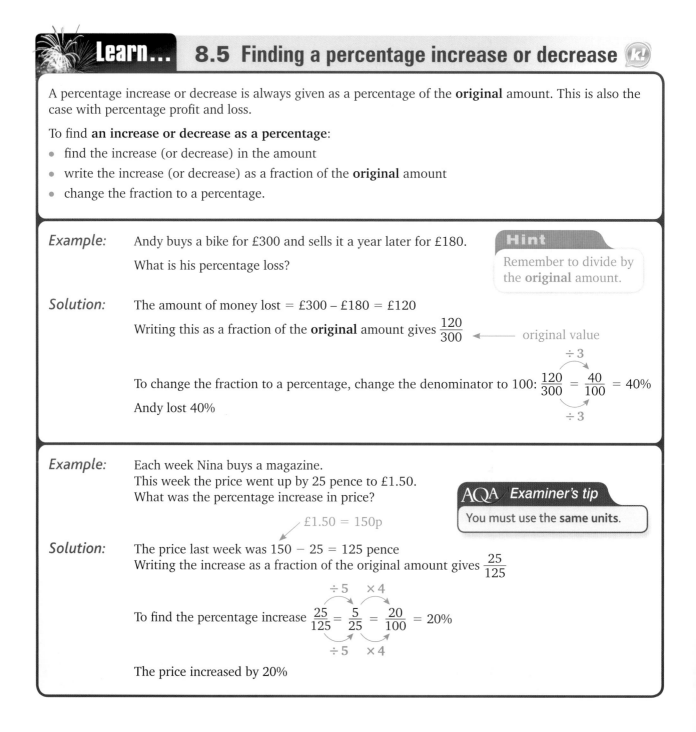

Learn... **8.5 Finding a percentage increase or decrease**

A percentage increase or decrease is always given as a percentage of the **original** amount. This is also the case with percentage profit and loss.

To find **an increase or decrease as a percentage**:

- find the increase (or decrease) in the amount
- write the increase (or decrease) as a fraction of the **original** amount
- change the fraction to a percentage.

Example: Andy buys a bike for £300 and sells it a year later for £180.
What is his percentage loss?

Hint
Remember to divide by the **original** amount.

Solution: The amount of money lost = £300 − £180 = £120

Writing this as a fraction of the **original** amount gives $\frac{120}{300}$ ← original value

To change the fraction to a percentage, change the denominator to 100: $\frac{120}{300} = \frac{40}{100} = 40\%$

Andy lost 40%

Example: Each week Nina buys a magazine.
This week the price went up by 25 pence to £1.50.
What was the percentage increase in price?

£1.50 = 150p

AQA *Examiner's tip*

You must use the **same units**.

Solution: The price last week was 150 − 25 = 125 pence
Writing the increase as a fraction of the original amount gives $\frac{25}{125}$

To find the percentage increase $\frac{25}{125} = \frac{5}{25} = \frac{20}{100} = 20\%$

The price increased by 20%

8.5 Finding a percentage increase or decrease

Practise...

G F E D C

C

1   The price of a bus fare goes up from 80 pence to £1.
    Find the percentage increase.

2   The number of fish in a pond has gone down from 200 to 180.
    What is the percentage decrease?

3   A day at a jet ski centre usually costs £150.
    If you take your own jet ski, it costs £120.
    What is the discount for taking your own jet ski as a
    percentage of the usual price?

4   Shona invests £480 in shares and sells them one year later for £600.
    What is the percentage increase?

5   Dylan buys a scooter for £4000.
    He sells it a year later for £2800.
    Find his percentage loss.

> **Hint**
> The percentage **loss** is the percentage **decrease** in price.

6   Sam's pay rate goes up from £7.00 to £7.35 per hour.
    Find the percentage increase.

7   A camera shop reduces the price of a digital camera from
    £60 to £40. Tracey says the price is reduced by 50%

    a   What mistake has she made?

    b   What is the actual percentage reduction?

> **Bump up your grade**
> For a Grade C you must be able to write an increase or decrease as a percentage.

8   The table shows the prices of a football club's season tickets.

    a   Work out which type of ticket has gone up by the
        greatest percentage.

    b   Each year Wayne buys a season ticket for himself
        and his daughter who is now 13 years old.
        Work out the percentage increase in the total cost of the tickets Wayne buys.

	Last year	This year
Adult	£250	£320
Under 16	£150	£180
Under 12	£90	£126

9   A sports shop sells tennis balls for 80 pence each or £2 for a pack of three.

    Marc buys a pack of three balls.
    Work out how much he saves as a percentage of the cost of buying the balls
    separately.

80p each or £2 for pack of 3

⚠ 10   A property developer buys some land for £1 million.
       He builds 25 houses on this land.
       It costs him £120 000 to build each house.
       He sells them for £200 000 each.
       Find the percentage profit.

> **Hint**
> The percentage profit is the percentage increase from the cost to the selling price.

11 Lisa says that it is not possible for something to increase by more than 100%
Do you agree? Explain your answer.

12 It costs a company £0.5 million to make 20 thousand computer games.

The company sells them at £30 each.

Find the percentage profit.

13 What is the percentage increase if the value of an antique:

   a increases by a quarter of its previous value

   b increases by one-third of its previous value

   c goes up to four times its previous value?

14 A supermarket sells crisps in a multi-pack of 12 packets for £2.16.
It also sells separate packets of crisps for 30 pence each.

   a What percentage do you save if you buy a multi-pack rather than separate packets of crisps?

   b Give two reasons why a shopper might decide to buy separate packets of crisps rather than a multi-pack.

15 What percentage extra do you get if you:

   a buy one, get one free

   b get three for the price of two

   c get an extra one free when you buy a pack of five?

Assess 🔊

1 **a** What percentage of this shape is shaded?

   b What percentage of the shape is not shaded?

   c Another shape has 40% shaded.
   What fraction of the shape is not shaded?

2 Copy and complete the table.
Write each fraction in its simplest form.

Decimal	0.3			0.05			1.5	
Fraction		$\frac{3}{4}$			$\frac{2}{5}$		$3\frac{1}{5}$	
Percentage			8%			32%		280%

G

F

3   Which of these values are less than $\frac{1}{3}$?   35%   0.0495   3.8% $\frac{3}{8}$ 0.329

4   Which is the larger amount?
You **must** show your working.

| 45% of £30 | | $\frac{4}{5}$ of £20 |

5   A watch priced at £70 is reduced by 20% in a sale.
What is the sale price?

6   A plumber charges £460 plus VAT charged at $17\frac{1}{2}$%. Work out the total bill.

7   81 out of 180 pupils in a school year are girls. What percentage is this?

8   Simon wants to buy a motorbike priced at £1500.
The dealer offers him two options.

| Option 1 – 20% deposit plus 12 monthly payments of £105. | Option 2 – Single cash payment with a discount of 10% |

a   How much more does it cost if Simon chooses Option 1 rather than Option 2?

b   Give one reason why Simon may still choose Option 1.

9   A bus fare goes up from 84 pence to £1.05.
What is the percentage increase?

10   A parents' association buys 50 Christmas trees for a total of £400 from a supplier to sell at a school's Christmas Sale.

a   They sell 35 of the trees for £12 each.
Work out the percentage profit on these trees.

b   At the end of the sale, they sell off the remaining trees for £5 each.
Work out the percentage loss on these trees.

c   What was the overall profit or loss?

AQA Examination-style questions

1   Three students take a test. The test has 80 questions.
Each question has 1 mark.
The pass mark is 75%

a   Did Chloe pass the test? Explain your answer.   (*1 mark*)

b   Did Tom pass the test? Explain your answer.   (*1 mark*)

c   How many questions did Rachel answer correctly?   (*1 mark*)

2   Andy's salary is £24 000 per year. He is paid the same amount each month.
He is given a pay rise of 10%
Calculate his new monthly salary.
You **must** show all your working.   (*4 marks*)

AQA 2008

9 Indices

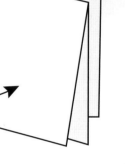

Objectives

Examiners would normally expect students who get these grades to be able to:

F

work out or know simple squares and square roots

E

work out or know simple cubes and cube roots

D

use the terms square, positive square root, negative square root, cube and cube root

recall integer squares from 2 × 2 to 15 × 15 and the corresponding square roots

recall the cubes of 1, 2, 3, 4, 5 and 10 and the corresponding cube roots

C

use index notation and index laws for multiplication and division for positive integer powers.

Did you know?

Folding paper

Did you know that it is impossible to fold a piece of paper more than 12 times?

If you fold the paper in half your paper is two sheets thick.

If you fold it in half again your paper is four sheets thick.

If you fold it in half again your paper is eight sheets thick.

How thick would your paper be after 12 folds?
Use the fact that paper is 0.1 millimetre thick.

Key terms

square number
cube number
square root
cube root
index
power
indices

You should already know:

✔ how to add, subtract, multiply and divide whole numbers

✔ how to use negative numbers

✔ how to use algebra.

Learn... 9.1 Powers and roots

Square numbers

A **square number** is the outcome when a number is multiplied by itself.

16 is a square number because $4 \times 4 = 16$ ←——— 4 squared

-4 squared is $-4 \times -4 = 16$

Hint

A negative number times a negative number is always a positive number.

A square number is a number 'to the power of 2' so 4 squared is also 4 to the power 2, which is written as 4^2.

Cube numbers

A **cube number** is the outcome when a number is multiplied by itself then multiplied by itself again.

125 is a cube number because $5 \times 5 \times 5 = 125$ ←——— 5 cubed

-5 cubed is $-5 \times -5 \times -5 = -125$

A cube number is a number 'to the power of 3' so 5 cubed is also 5 to the power 3, which is written as 5^3.

Square roots

The **square root** of a number, such as 16, is the number that gives 16 when multiplied by itself.

The square root of 16 is 4 because $4 \times 4 = 16$

However, the square root of 16 is also -4 because $-4 \times -4 = 16$

The square root of 16 is written as $\sqrt{16}$ or $\sqrt[2]{16}$, so $\sqrt{16} = 4$ (or -4)

Cube root

The **cube root** of a number, such as 125, is the number that gives 125 when multiplied by itself then multiplied again.

The cube root of 125 is 5 because $5 \times 5 \times 5 = 125$

The cube root of 125 is written as $\sqrt[3]{125}$, so $\sqrt[3]{125} = 5$

AQA Examiner's tip

Do not confuse cube roots with square roots where you have two answers. The $\sqrt[3]{125}$ is not -5 because $-5 \times -5 \times -5 = -125$

Practise... 9.1 Powers and roots k!

G F E D C

1 Write down the value of:

a 6^2　　c 13^2　　e $\sqrt{121}$　　g 1^2　　i $\sqrt{225}$

b 9^2　　d $\sqrt{25}$　　f $\sqrt{64}$　　h $\sqrt{81}$

2 　3　5　8　12　15　20　25

From the list of numbers above, write down:

a a square number

b a cube number

c the square root of 144

d the cube root of 512.

3 Write down the value of:

a 2^3　　b 4^3　　c 10^3　　d $\sqrt[3]{216}$　　e $\sqrt[3]{1}$　　f $\sqrt[3]{27}$

4 Work out:

a $1^3 + 4^2$　　d $\sqrt[3]{8} + 4^2$　　g $\sqrt[3]{1\,000\,000} - \sqrt{10\,000}$

b $6^2 - 3^2$　　e $\sqrt[3]{1000} - \sqrt{81}$

c $\sqrt{144} - \sqrt{100}$　　f $\sqrt{169} - 3^2$

F

F
E

E

E

5 Neil says -5^2 is 25

Andrea says -5^2 is -25

Who is correct?

Give a reason for your answer.

E
D

6 Which is the higher number?

a 2^3 or 3^2　　　**b** $\sqrt{64}$ or $\sqrt[3]{125}$　　　**c** $\sqrt[3]{-8}$ or $-\sqrt{9}$?

D

7 Write down an approximate answer to the following.

a 4.99^2　　　**b** $\sqrt{50}$　　　**c** $\sqrt[3]{999}$

8 Vivek says that $-11^2 = 121$

Is this correct?

Give a reason for your answer.

> **Bump up your grade**
>
> To get a Grade C you need to be able to find cube roots of negative numbers and know that square numbers have a positive and a negative square root.

! 9 Write down the square roots of 0.01

! 10 Write down the cube root of -0.027

? 11 The number 50 can be written as $5^2 + 5^2$ or as $1^2 + 7^2$

a Write 100 as the sum of square numbers in as many different ways as you can.

b Can you write the number 100 as the sum of cube numbers?

? 12 Jenny investigates the sum of the cubes of the first two integers.

She notices that the sum gives a square number:
$1^3 + 2^3 = 9 \ (= 3^2)$

Jenny now investigates the sum of the cubes of the first three integers.

She notices, again, that the sum gives a square number:
$1^3 + 2^3 + 3^3 = 36 \ (= 6^2)$

Does this work for the sum of the cubes of the first four integers?

Give a reason for your answer.

What about the other sums of consecutive cubes?

Learn... 9.2 Rules of indices 🔊

The **index** (or **power**) tells you how many times the base number is to be multiplied by itself.
This means that 10^3 tells you that 10 (the base number) is to be multiplied by itself three times (the index or power).

index (or power)

$$10^3$$

base

So $10^3 = 10 \times 10 \times 10 = 1000$

Rules of indices

$a^3 \times a^5 = (a \times a \times a) \times (a \times a \times a \times a \times a)$ $= a \times a \times a \times a \times a \times a \times a \times a$ $= a^8$	So $a^3 \times a^5 = a^8$ In general $a^m \times a^n = a^{m+n}$
$a^7 \div a^3 = \dfrac{a^7}{a^3}$ $= \dfrac{a \times a \times a \times a \times a \times a \times a}{a \times a \times a}$ $= \dfrac{\not{a} \times \not{a} \times \not{a} \times a \times a \times a \times a}{\not{a} \times \not{a} \times \not{a}}$ $= a \times a \times a \times a$ $= a^4$	So $a^7 \div a^3 = a^4$ In general $a^m \div a^n = a^{m-n}$
$(a^2)^3 = a^2 \times a^2 \times a^2$ $= (a \times a) \times (a \times a) \times (a \times a)$ $= a \times a \times a \times a \times a \times a$ $= a^6$	So $(a^2)^3 = a^6$ In general $(a^m)^n = a^{m \times n}$

Example: Simplify:

	Number	*Algebra*
a	$6^3 \times 6^2$	$a^3 \times a^2$
b	$\dfrac{2^5}{2^2}$	$\dfrac{a^5}{a^2}$
c	$(3^5)^2$	$(a^5)^2$

Solution:

	Number	*Algebra*
a	$6^3 \times 6^2$ $= 6^{(3+2)}$ $= 6^5$	$a^3 \times a^2$ $= a^{(3+2)}$ $= a^5$
b	$\dfrac{2^5}{2^2}$ $= 2^5 \div 2^2$ $= 2^{(5-2)}$ $= 2^3$	$\dfrac{a^5}{a^2}$ $= a^5 \div a^2$ $= a^{(5-2)}$ $= a^3$
c	$(3^5)^2$ $= 3^{5 \times 2}$ $= 3^{10}$	$(a^5)^2$ $= a^{5 \times 2}$ $= a^{10}$

Practise... 9.2 **Rules of indices** k!   G F E D C

1 Write in index notation:

a $5 \times 5 \times 5 \times 5$

b $10 \times 10 \times 10 \times 10 \times 10 \times 10 \times 10$

c $6 \times 6 \times 6 \times 6 \times 6 \times 6 \times 6 \times 6 \times 6 \times 6 \times 6 \times 6$

d 13×13

e $2 \times 2 \times 2 \times 2 \times 2 \times 2 \times 2 \times 2 \times 2 \times 2 \times 2 \times 2 \times 2 \times 2 \times 2 \times 2 \times 2 \times 2$

f $12 \times 12 \times 12 \times 12$

g $p \times p \times p \times p$

h $s \times s \times s \times s \times s \times s \times s \times s \times s \times s$

2 Work out the value of each of the following.

a 7^2 c 11^2 e 2^3 g 1^5 i 4^3

b 4^2 d $(-3)^2$ f 10^4 h 3^4 j $(-10)^6$

3 Use the rules of **indices** to simplify the following. Give your answers in index form.

a $5^6 \times 5^2$ e $10^6 \times 10^{12}$ i $\dfrac{9^{12}}{9^{11}}$

b $12^8 \times 12^3$ f $7^{11} \div 7^6$ j $(6^2)^5$

c $3^5 \div 3^2$ g $6^5 \div 6^3$ k $(11^5)^4$

d $4^3 \times 4^8$ h $\dfrac{4^7}{4^3}$ l $(10^{10})^{10}$

4 Are the following statements true or false? Give a reason for your answer.

a $6^2 = 12$ b $1^3 = 1$ c $\dfrac{2^{10}}{4^5} = 1$ d $3^2 + 3^3 = 3^5$

5 Simplify the following.

a $x^6 \times x^2$ c $\dfrac{a^7}{a^3}$ e $q^7 \div q^7$

b $e^8 \times e^3$ d $p^{10} \div p^5$ f $(b^2)^5$

6 Adnan writes: $a^2 \div a^2 = \dfrac{a \times a}{a \times a}$

$$= \dfrac{\not{a} \times \not{a}}{\not{a} \times \not{a}} = \dfrac{1}{1} = 1$$

He says that $a^0 = 1$

Is he correct?

Give a reason for your answer.

7 The number one million $= 10^6$ which is $10 \times 10 \times 10 \times 10 \times 10 \times 10 = 1\,000\,000$

Write down the value of: a   one billion $= 10^9$

b   one trillion $= 10^{12}$

8 The number 64 can be written as 8^2 in index form.

Write down three other ways of writing 64 in index form.

9 Sue says the sum of the squares of two odd numbers is always odd.

Ravi says the sum of the squares of two odd numbers is always even.

Keith says the sum of the squares of two odd numbers could be odd or even.

Who is correct? Give a reason for your answer.

9 Assess 🄺!

1 Evaluate the following.

   a 4^2 **b** 11^2 **c** $\sqrt{36}$ **d** $\sqrt{196}$ **e** 0^2 **f** $(-3)^2$

2 Evaluate the following.

   a 5^3 **b** 10^3 **c** $\sqrt[3]{27}$ **d** $\sqrt[3]{64}$ **e** $\sqrt[3]{0}$ **f** $\sqrt[3]{-8}$

3 Write down an approximate answer to the following.

   a 9.99^2 **b** $\sqrt{102}$ **c** $\sqrt[3]{-126}$

4 **a** Sam says all numbers have two square roots. Gareth says some numbers have no square roots. Who is right?

   Give a reason for your answer.

   b Livia joins in the conversation and says that all numbers have two cube roots. Is she right?

   Give a reason for your answer.

5 Simplify the following, leaving your answers as single powers.

   a $4^6 \times 4^2$ **d** $7^5 \times 7$ **g** $21^7 \div 21^5$ **j** $2^3 \div 2^3$

   b $11^5 \times 11^3$ **e** $6^4 \times 6^2 \times 6^3$ **h** $16^{10} \div 16^9$

   c $(5^3)^2$ **f** $10^4 \div 10^2$ **i** $5^8 \div 5^7$

6 Find the value of:

   a $3^2 \times 4^2$ **b** $5^3 \div 5^2$ **c** $6^5 \times 6^3 \div 6^4$ **d** $\dfrac{10^8 \times 10^7}{10^7 \times 10^6}$

7 Which is greater:

   a 3^5 or 5^3 **b** 11^2 or 2^{11} **c** 2^4 or 4^2 ?

8 Alex says the sum of two consecutive squares is always odd.

   Kate says the sum of two consecutive squares is always even.

   Dan says the sum of two consecutive squares could be odd or even.

   Who is correct? Give a reason for your answer.

AQA Examination-style questions 🄺!

1 **a** Simplify $t^4 \times t^5$ *(1 mark)*

   b Simplify $p^6 \div p^2$ *(1 mark)*

   c **i** Chris simplifies 2×2^5
   His answer is 2^5
   Explain the mistake he has made. *(1 mark)*

   ii Simplify $3^6 \div 3$
   Write your answer as a power of 3. *(1 mark)*

AQA 2008

10 Graphs of linear functions

10

Objectives

Examiners would normally expect students who get these grades to be able to:

E

produce a table of values for equations such as $y = 3x - 5$ or $x + y = 7$ and draw their graphs

D

solve problems such as finding where the line $y = 3x - 5$ crosses the line $y = 4$

C

find the gradients of straight-line graphs.

Did you know?

Rollercoaster

The design of a rollercoaster has to have a long slope with a chain lift to drag the rollercoaster car to the top. This gives it enough energy to reach the end of the track. The slope needs to be high enough so that the car will roll along to the end of the track. The designer has to choose a gradient for the first slope. If it is too steep, it could be unsafe. If it is too shallow, the rollercoaster ride would take up too much space in the theme park. Choosing the proper gradient is very important.

Key terms

linear
gradient
variable
coefficient

You should already know:

✔ how to plot points in all four quadrants

✔ how to recognise lines such as $y = 4$ or $x = -3$

Learn... 10.1 Drawing straight-line graphs 🎵

An equation such as $y = 3x - 5$ can be shown on a graph.

The graph will be a straight line and $y = 3x - 5$ is called a **linear** equation.

A linear equation does not contain any powers of x or y.

To draw the graph, you need to work out the coordinates of three points on the line.

You may be given a table of values to use.

If not, choose any three values of x that lie within the range you have been given.

Work out the corresponding y values, using the linear equation.

Plot the points. Draw the line through the points.
The line must go across the full range of values for x.

> **AQA Examiner's tip**
>
> It is a good idea to use zero as one of your x-values, as it is easy to substitute.

Example: Draw the graph of $y = 3x - 5$ for values of x from -2 to 4.

Solution: Choose three values, for example $x = 0$ and $x = -2$ and $x = 4$, the end values.

Work out the y values and put them in a table.

> **AQA Examiner's tip**
>
> Always use three points, not just two. They should be in a straight line. If they are not, you have made a mistake in working out one of the y-values, so check and correct the values.

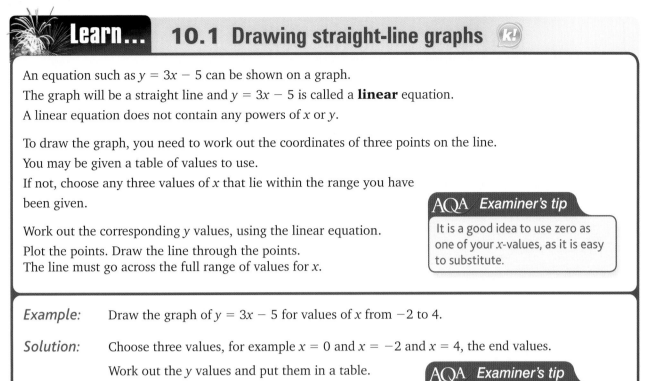

Plot the points.

(In the exam the axes will be drawn for you.)

Draw the line through the plotted points, making sure it goes from $x = -2$ to $x = -4$

This straight-line graph has been drawn using the same scale on both axes.

This makes it easy to plot the points.

The range of y-values is large, so the graph is tall and narrow.

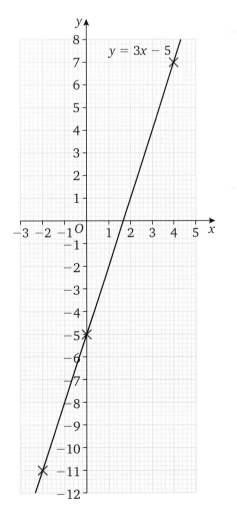

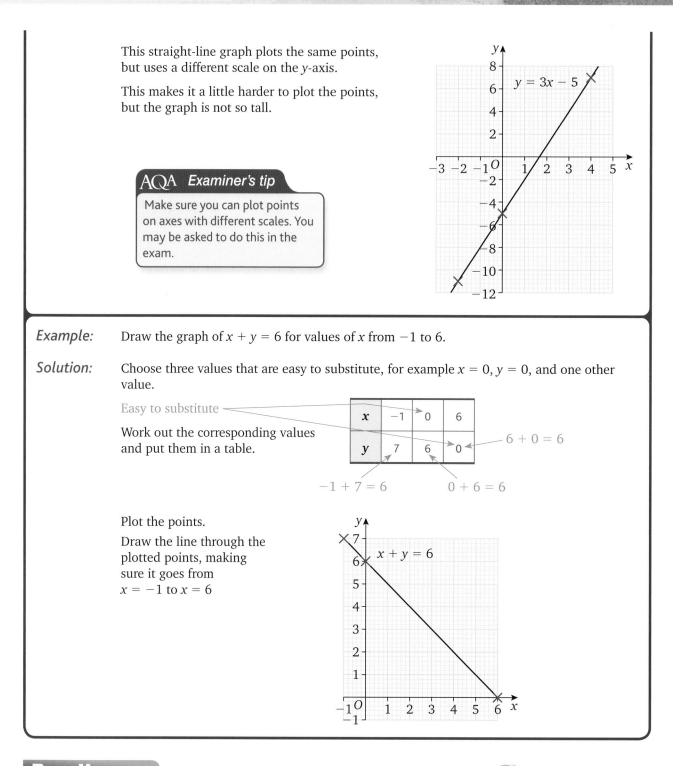

This straight-line graph plots the same points, but uses a different scale on the *y*-axis.

This makes it a little harder to plot the points, but the graph is not so tall.

AQA *Examiner's tip*

Make sure you can plot points on axes with different scales. You may be asked to do this in the exam.

$y = 3x - 5$

Example: Draw the graph of $x + y = 6$ for values of *x* from -1 to 6.

Solution: Choose three values that are easy to substitute, for example $x = 0, y = 0$, and one other value.

Easy to substitute

Work out the corresponding values and put them in a table.

x	-1	0	6
y	7	6	0

$6 + 0 = 6$

$-1 + 7 = 6$ $0 + 6 = 6$

Plot the points.

Draw the line through the plotted points, making sure it goes from $x = -1$ to $x = 6$

$x + y = 6$

Practise... **10.1 Drawing straight-line graphs** (k!) G F E D C

Next to each of the first three questions there is a sketch to show you the range you will need on your axes.
The sketch is not drawn to scale.

1 **a** Draw the graph of $y = x + 2$ for values of *x* from -3 to 4.

b Write down the coordinates of the point where this graph crosses the *y*-axis.

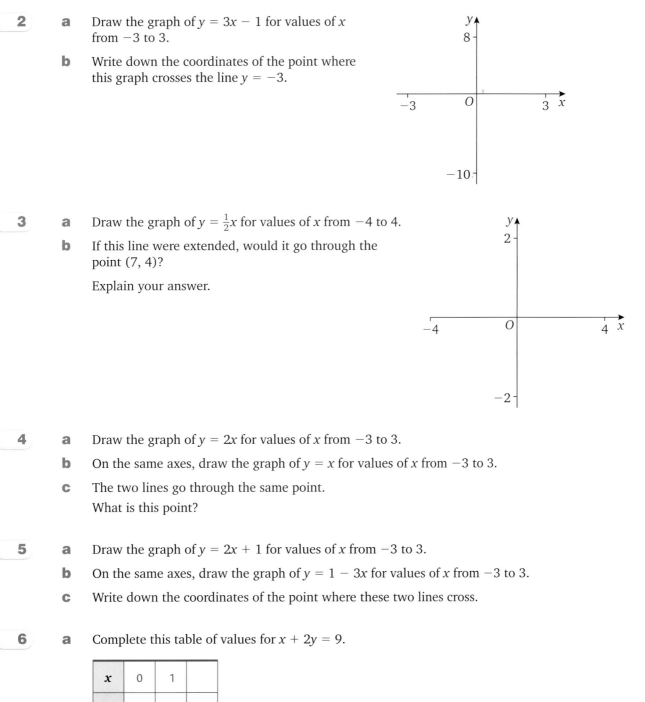

2   **a**   Draw the graph of $y = 3x - 1$ for values of x from -3 to 3.

b   Write down the coordinates of the point where this graph crosses the line $y = -3$.

3   **a**   Draw the graph of $y = \frac{1}{2}x$ for values of x from -4 to 4.

b   If this line were extended, would it go through the point $(7, 4)$?

Explain your answer.

4   **a**   Draw the graph of $y = 2x$ for values of x from -3 to 3.

b   On the same axes, draw the graph of $y = x$ for values of x from -3 to 3.

c   The two lines go through the same point.
What is this point?

5   **a**   Draw the graph of $y = 2x + 1$ for values of x from -3 to 3.

b   On the same axes, draw the graph of $y = 1 - 3x$ for values of x from -3 to 3.

c   Write down the coordinates of the point where these two lines cross.

6   **a**   Complete this table of values for $x + 2y = 9$.

x	0	1	
y		0	

b   Draw the graph of $x + 2y = 9$ for values of x from 0 to 9.

c   Write down the coordinates of the point where your graph crosses the line $x = 4$.

7   **a**   Complete this table of values for $x - 2y = 1$.

x	0		3
y		0	

b   Draw the graph of $x - 2y = 1$ for values of x from -3 to 3.

c   Write down the coordinates of the point where your graph crosses the line $y = \frac{1}{2}$.

8   Which of these equations represent straight-line graphs?

A $y = 1 - 8x$ **B** $2y = 5x + 4$ **C** $y = x^2 + 7$ **D** $4x + 3y = 8$

9   Which of these points lies on the line $3x + 2y = 12$? Show how you found your answers.

A $(0, 4)$ D $(1, 4\frac{1}{2})$

B $(2, 3)$ E $(6, -3)$

C $(3, 2)$ F $(-2, 8)$

10 $P(-3, 6)$, $Q(0, 0)$ and $R(2, -4)$ are three points on a straight line.

Which of the following is the equation of the line?

A $y = x + 9$ B $x + y = 3$ C $y + 2x = 0$

Show how you found your answer.

11   Each of the following points lies on one or more of the given lines.

Match the points to their lines.

Points: $A(-2, 7)$ $B(0, 0)$ $C(1, 4)$ $D(2, 5)$ $E(3, 3)$ $F(4, 1)$

Lines: $y = 4x$ $2x + y = 9$ $x + y = 5$ $y = 6x - 7$

Learn... 10.2 Gradients of straight-line graphs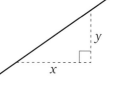

The **gradient** of a straight-line graph is a measure of how steep it is.

A line that slopes from top right to bottom left has a positive gradient, because y increases as x increases.

The gradient can be found from the graph of the line.

$$\text{Gradient} = \frac{\text{change in vertical distance}}{\text{change in horizontal distance}} = \frac{y}{x}$$

To find the gradient, draw a line parallel to the x-axis and a line parallel to the y-axis to make a right-angled triangle on the graph.

The triangle can be anywhere on the graph.

Example: Find the gradient of the graph below.

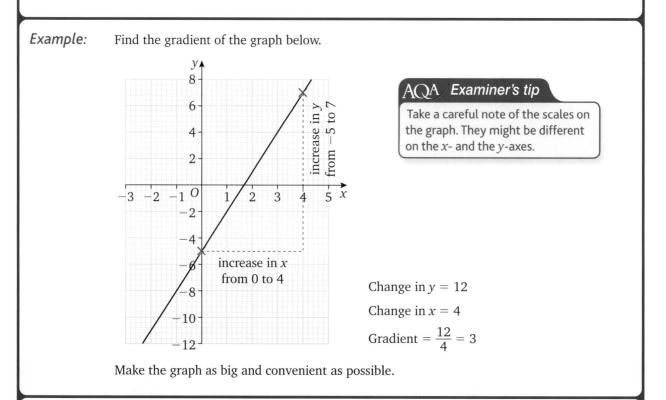

AQA *Examiner's tip*

Take a careful note of the scales on the graph. They might be different on the x- and the y-axes.

increase in y from -5 to 7

increase in x from 0 to 4

Change in $y = 12$

Change in $x = 4$

Gradient $= \frac{12}{4} = 3$

Make the graph as big and convenient as possible.

A line that slopes from top left to bottom right has a negative gradient because y decreases as x increases.

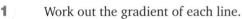

$$\text{Gradient} = -\frac{y}{x}$$

The gradient can also be found from the equation of the line.

To find the gradient, write the equation of the line in the form: $y = mx + c$

y and x are the **variables** in the equation.

m (the **coefficient** of x) is the gradient of the line.

Example: What is the gradient of $y = 5x + 2$?

Solution: The coefficient of x is 5, so the gradient is 5.

Example: What is the gradient of $y = 3 - 2x$?

Solution: The coefficient of x is -2, so the gradient is -2.

Example: What is the gradient of $x + y = 5$?

Solution: Make sure the equation is in the form $y = mx + c$

$$x + y = 5$$
$$x + y - x = 5 - x \qquad \text{Subtract } x \text{ from both sides.}$$
$$y = 5 - x$$

The coefficient of x is -1 so the gradient is -1.

> **Bump up your grade**
>
> To get a Grade C you need to be able to find the gradient of straight-line graphs.

10.2 Gradients of straight-line graphs

Practise...

G   F   E   D   C

1   Work out the gradient of each line.

a

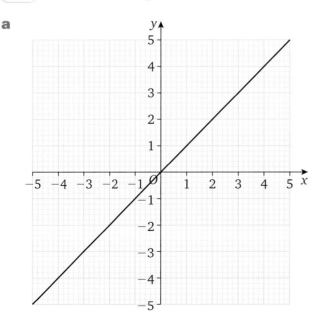

b

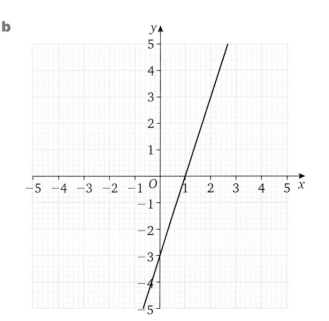

C

c

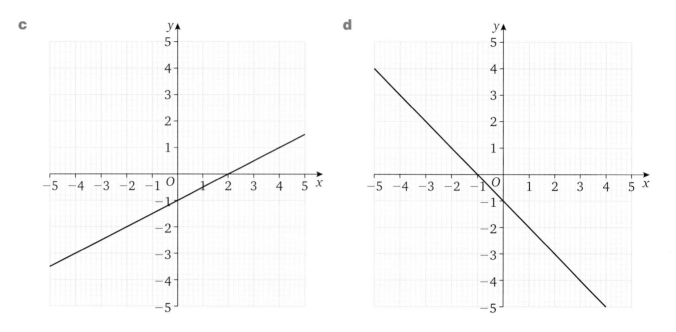

2 Work out the gradient of each of these straight lines.

a $y = 5x + 4$

b $y = 2 + x$

c $y = 3 - 2x$

d $y + 5 = 3x$

e $2y = 6x - 7$

f $4x + y = 9$

3 Jo says that the lines whose equations are $y = 5 - 2x$ and $y = 5 - 4x$ have the same gradient. Explain why Jo is wrong.

4 The diagram opposite shows four lines labelled A, B, C, D.

a Which line has a gradient of 2?

How do you know?

b Which line has a gradient of 1?

How do you know?

c Which line or lines have a negative gradient?

How do you know?

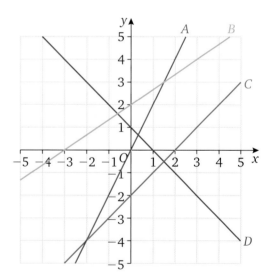

5 The diagram opposite shows the line $y = 3x - 5$

$RQ = 3$ units.

What is the length of PQ?

Show your working.

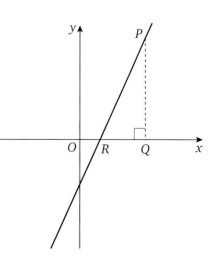

! **6**   On the same axes, draw the graphs of $y = 2x$, $y = 2x + 4$ and $y = 2x - 5$ for values of x from -4 to 4.

What do you notice?
How does this relate to the equations?

7   Use your knowledge of gradients to match the equations to the sketch graphs.

$y = 3x$ $y = -2x$ $y = 4 - x$ $y = 3x + 8$

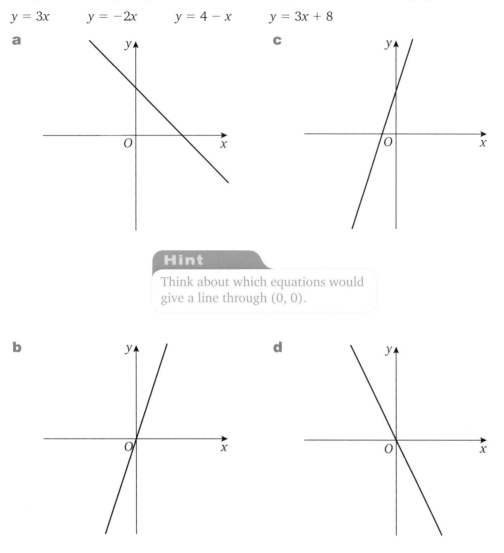

a

c

Hint

Think about which equations would give a line through $(0, 0)$.

b

d

Assess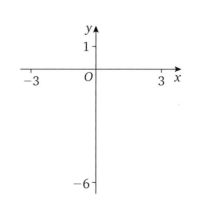

1   **a**   Draw the graph of $y = x - 3$ for values of x from -3 to 4.

b   Write down the coordinates of the point where this graph crosses the y-axis.

E
D

2　**a**　Draw the graph of $y = 5 - 4x$ for values of x from -2 to 3.

　　b　Write down the coordinates of the point where this graph crosses the line $y = 2$

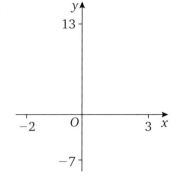

3　**a**　Complete this table of values for $2x + y = 8$

x	-1	0	5
y			-2

　　b　Draw the graph of $2x + y = 8$ for values of x from -1 to 5.

　　c　Write down the coordinates of the point where this graph crosses the line $y = 3$

4　**a**　Complete this table of values for $y = 3 + \frac{1}{2}x$

x	-2	0	4
y			5

　　b　Draw the graph of $y = 3 + \frac{1}{2}x$ for values of x from -2 to 4.

　　c　If this graph were extended, would it go through the point $(6, 6)$?
　　　　Explain your answer.

D

5　Which of these points lie on the line $4x - 3y = 4$?

　　A　$(0, 1)$　　　　　**C**　$(3, 4)$　　　　　**E**　$(7, 8)$

　　B　$(1, 0)$　　　　　**D**　$(4, 4)$　　　　　**F**　$(8, 7)$

　　Show how you found your answers.

C

6　Work out the gradient of each of these lines.

　　a　$y = 5x - 1$　　　**b**　$y = 9 - 2x$　　　**c**　$3x + y = 2$　　　**d**　$y - x = 3$

7　Which of these equations does not represent a straight-line graph?

　　A　$3x + 5y + 1 = 0$　　**B**　$y^2 = 2x - 5$　　　**C**　$y = 12 - x$　　　**D**　$4 = 4y + 3x$

8　The gradient of this line is -2.

　　$BC = 4$ units

　　What is the length of AB?

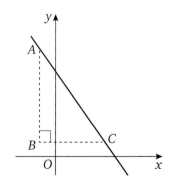

9   Jacqui says the equation of this graph is $y = 3x + 4$.

Explain how you can tell, by looking at the graph, that she is wrong.

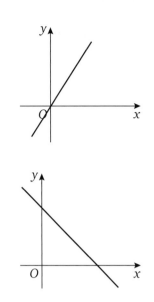

C

10   Rasheed says the equation of this graph is $y = x - 5$.

Ben says it is $y = 5 - x$.

Look at the graph to decide who is wrong and explain how you made your decision.

AQA Examination-style questions

1   The graph shows a sketch of the line $y = 3x + 1$

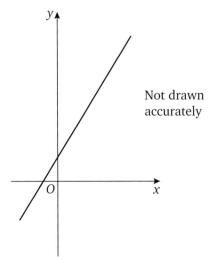

Not drawn accurately

a   Does the point $(-2, -5)$ lie on the line?
Explain your answer.   *(2 marks)*

b   Copy the graph and sketch the line $y = 3x + 4$ *(2 marks)*

AQA 2008

Objectives

Examiners would normally expect students who get these grades to be able to:

G

use a formula in words such as:
Total pay = rate per hour × no. of hours + bonus

F

substitute positive numbers into a simple formula such as $P = 2L + 2W$

derive simple expressions

E

use formulae such as $P = 2L + 2W$ to find W given P and L

substitute negative numbers into a simple formula

use formulae from mathematics and other subjects

derive expressions and formulae

D

substitute numbers into more complicated formulae such as $C = \dfrac{(A + 1)}{9}$

derive more complex expressions and formulae

distinguish between an expression, an equation and a formula

C

rearrange linear formulae such as $p = 3q + 5$

I used the formula: cooking time = 20 per pound weight + 20 to work out how long the chicken needed. Perhaps the time should have been in minutes not hours!

Did you know?

Formulae can be very useful

Formulae are used in everyday life, for example in cooking instructions. Be careful to check that the units you are working with make sense!

Key terms

expression
formula
symbol
substitute
values
equation
subject

You should already know:

✔ order of operations (BIDMAS)

✔ the four rules applied to negative numbers

✔ how to simplify expressions by collecting like terms

✔ how to solve linear equations.

Learn... 11.1 Writing formulae using letters and symbols

An **expression** is a collection of algebraic terms such as $3x - 2y + z$.

A **formula** tells you how to work something out such as $A = l \times w$

This formula tells you to multiply l and w to obtain A.

A formula can be written using words or **symbols**.

When you write expressions and formulae using symbols you need to know:

- If a stands for a number, then $2 \times a$ can be written as $2a$.
  Always write the number in front of the letter.

- The expression $3x + 5$ means multiply x by 3 and then add 5.

- The expression $5(x - 2)$ means subtract 2 from x and then multiply the answer by 5.

Example: A rough rule for changing inches to centimetres is to multiply the number of inches by 2.5
Write a formula for this rule.

Solution: In words the formula would be:

'number of centimetres' = 'number of inches' $\times$ 2.5

Choose letters to stand for 'number of centimetres' and 'number of inches'.

You could choose C for the 'number of centimetres' and i for the 'number of inches'.

$C = i \times 2.5$

$C = 2.5i$ *Remember to put the number in front of the letter.*

> **AQA** *Examiner's tip*
>
> Be careful when you choose your own letters in problems. Some letters are easily confused with numbers.
>
> Z and 2 can get confused.
> I and 1 can get confused.
> b and 6 can get confused.
> q and 9 can get confused.
> S and 5 can get confused.

Example: **1**   Write an expression for each of the following:

     a   Think of a number. Double it. Add 3.

     b   Think of a number. Add 1. Multiply by 2.

     2   Describe how the following expression is formed.

$$\frac{(x + 3)}{4} - 5$$

Solution: **1**   **a**   Think of a number, call it x x

          Double it $2x$ (*double* is the same as *multiply by 2*)

          Add 3 $2x + 3$

       b   Think of a number x

          Add 1 $x + 1$

          Multiply by 2 $2(x + 1)$

     2   Think of a number x

        Add 3 $x + 3$

       Divide by 4 $\dfrac{(x + 3)}{4}$

       Subtract 5 $\dfrac{(x + 3)}{4} - 5$

Example: Andy has *k* marbles.

     a   Ben has five more marbles than Andy.
        Write down an expression for the number of
        marbles Ben has.

     b   Chris has one fewer marble than Andy.
        Write down an expression for the number of
        marbles Chris has.

     c   Dean has twice as many marbles as Andy.
        Write down an expression for the number of marbles Dean has.

     d   Enid has three times as many marbles as Ben.
        Write down an expression for the number of marbles Enid has.

Solution: **a**   **Five more than** is the same as **add 5**.
        Ben has $k + 5$ marbles.

     b   **One fewer than** is the same as **subtract 1**.
        Chris has $k - 1$ marbles.

     c   **Twice as many** is the same as **multiply by 2**.
        Dean has $2k$ marbles.

     d   **Three times as many** is the same as **multiply by 3**.
        Enid has $3(k + 5)$ marbles.

> **AQA**   *Examiner's tip*
>
> Put brackets around expressions
> when you have to multiply them.
> $k + 5 \times 3$ would have given you
> the wrong number of marbles
> for Enid.

Example: Write a problem which leads to the expression $2x + 3$ as the answer.

Solution: There are *x* pencils in a pack.
          Lee has two packs and three extra pencils.
          Write an expression for the total number of pencils.
          Answer: $2x + 3$

Practise...

11.1 Writing formulae using letters and symbols

   **G F E D C**

F

1   Dermot has *x* books on his shelf.

     a   Ewan has five more books than Dermot.
        Write down an expression for the number of books Ewan has.

     b   Fred has three times as many books as Dermot.
        Write down an expression for the number of books Fred has.

2   Ranee earns £*y* per week.

     a   Sue earns £4 less than Ranee.
        Write down an expression for the amount Sue earns.

     b   Tina earns twice as much as Ranee.
        Write down an expression for the amount Tina earns.

3   Jon buys 12 bags of crisps.
       The price of a bag of crisps is *z* pence.
       Write down an expression for the total cost of the crisps.

4   Anna buys *m* apples.

Each apple costs 18p.

Write down an expression for the total cost of the apples.

5   A cookery book gives this rule for roast chicken:

> 40 minutes per kilogram plus 20 minutes

Write an expression for this rule.

Use *k* to stand for the number of kilograms.

6   Write an expression for each of these statements.

a   Four times the number *x*

b   Five times the number *z* plus six times the number *x*

c   The sum of two times the number *x* and double the number *y*

d   The product of the numbers *x*, *y* and *z*

e   The sum of seven times the number *x* and double the number *y*

f   Half the number *y* subtracted from six times the number *x*

7   Write an expression for each of these statements.

Let the unknown number be *x*.

a   Think of a number. Add 3.   **d**   Think of a number. Multiply by 3. Add 2.

b   Think of a number. Double it.   **e**   Think of a number. Add 1. Multiply by 2.

c   Think of a number. Subtract 5.

8   The rule for finding out how far away thunderstorms are is:

'Count the number of seconds between the lightning and the thunder.

Divide the answer by 5. The answer gives the distance in miles.'

Write a formula for this rule.

Use *s* to stand for the number of seconds and *d* to stand for the distance in miles.

9   You can find out how many amps an electrical appliance will use by using this rule:

Number of amps equals the number of watts divided by 240.

Write a formula for this rule.

Use *a* to stand for the number of amps and *w* to stand for the number of watts.

10   Write a problem which leads to the expression $3x + 2$ as the answer.

11   In Gordon's takeaway a pizza costs £*m* and a pasta dish costs £*n*.

On Thursday evening, he sells 32 pizzas and 15 pasta dishes.

Write down an expression for his total takings from pizza and pasta dishes on Thursday evening.

12   Jan sells homemade cakes to raise funds for her son's school.

She charges £3 for each cake but has to pay £5 for her stall.

She sells *n* cakes.

Write down an expression in *n* for the profit she makes.

D

13 Write down a formula for the total cost (*T*) of:

a *x* lollies at 70p each and *y* lollies at 80p each

b *c* cakes at 90p each and *b* biscuits at 20p each.

14 Phil bought *x* cups of tea and *y* cups of coffee from his local cafe for himself and his friends. The tea cost 80p per cup, and the coffee cost 90p per cup.

Which one of the following could be a formula for the total cost in pence, *C*?

$C = 90x + 80y$

$C = 80x + 90y$

$C = 80x90y$

Explain your answer, and state why the other formulae are not correct.

C

15 Sean and Dee organise a quiz evening.

A quiz team is made up of four people.

Each team is charged £8 to enter the quiz.

Sean and Dee spend £20 on prizes and £25 on food and drink.

There are *x* teams at the quiz evening.

On average, each team member spends £3 on refreshments.

Write down an expression in *x* for the profit they make.

16 Kate has £2 to spend. She buys *x* pens at *y* pence each.

a Write down an expression for the total amount she spends.

b Write down a formula for the amount of change she receives. Use *C* to stand for her change in pence.

AQA *Examiner's tip*

Make sure that units are consistent. If you are working in pounds then make sure all prices/costs are in pounds, if you are working in kilograms, then make sure that all weights are in kilograms, and so on.

17 When Mrs Tyler does group work in her classroom, she arranges the tables and chairs in the following way.

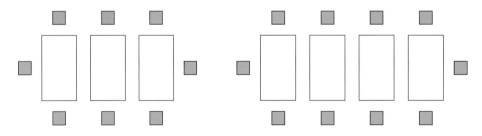

The first arrangement has three tables and eight chairs.

The second arrangement has four tables and ten chairs.

a How many chairs will there be if there are five tables arranged in this way?

b How many chairs will there be if there are ten tables arranged in this way?

c Copy and complete the following table for the number of tables and chairs.

Number of tables	1	2	3	4	5
Number of chairs					

d Write down a formula showing how to work out the number of chairs from the number of tables. Use the letter *T* for the number of tables and *C* for the number of chairs.

18   Natland Taxis use the formula $C = 3m + 4.5$ to work out the cost of journeys for customers.

C is the total charge in pounds and m is the number of the miles for the journey.
Every journey has a minimum charge and a charge for each mile.

  a   What is Natland Taxis' minimum charge?

  b   What is the cost per mile?

19   Sedgwick Tool Hire charge £18 to hire a cement mixer for one day. They charge £9 for every day extra.

Harry says the formula for the total charge is $C = 18d + 9$.

He uses d for the number of days and C for the total charge.

  a   Explain why Harry's formula is not correct.

  b   Write down what Harry's formula should be.

  c   Use the internet to find some costs for tool hire. Write down some formulae for the cost of tool hire.

Learn...   11.2 Using formulae

When you use a formula you **substitute** the **values** you are given into it. This means you replace the letters with the numbers given to you.

Start by writing the expression or formula down, and then substitute the numbers given.

When you substitute numbers into a formula you get an equation.
This equation can then be solved to find the value of the unknown letter.

Example: If $a = 2$, $b = 3$, $c = -1$, find the value of y when

  a $y = a + b + c$

  b $y = ab$

  c $y = 4a - 3b + 2c$

  d $y = \dfrac{3a}{2b}$

Solution: **a** $y = a + b + c$ Write the expression down.

  $y = 2 + 3 - 1$ Substitute the numbers given.

  $y = 5 - 1$

  $y = 4$

  b $y = ab$ Write the expression down.

  $y = 2 \times 3$ Substitute the numbers given (remember ab is the same as $a \times b$).

  $y = 6$

  c $y = 4a - 3b + 2c$ Write the expression down.

  $y = (4 \times 2) - (3 \times 3) + (2 \times -1)$ Substitute the numbers given.

  $y = 8 - 9 - 2$ Take care with signs (remember $2 \times -1 = -2$).

  $y = -3$

  d $y = \dfrac{3a}{2b}$ Write the expression down.

  $y = \dfrac{3 \times 2}{2 \times 3}$ Substitute the numbers given.

  $y = \dfrac{6}{6}$

  $y = 1$

> **AQA** *Examiner's tip*
>
> You can use brackets to help you do calculations in the correct order.
>
> Remember, BIDMAS applies to algebra as well as arithmetic.

Example: Maxi is using the formula $p = 3x + 5xy$

He needs to find p when $x = 2$ and $y = 3$

Solution:

$p = 3x + 5xy$	Write the formula down.
$p = (3 \times 2) + (5 \times 2 \times 3)$	Substitute the numbers given.
$p = 6 + 30$	Take care to work this out in the correct order.
$p = 36$	

AQA *Examiner's tip*

Remember to show all stages of your working; you gain method marks for this in an examination.

Practise... 11.2 Using formulae   G F E D C

1   Tasty Pizzas works out its delivery charge with the formula:
Delivery Charge (£) = Number of pizzas $\times$ 0.50 + 1.25
Find the delivery charge for three pizzas.

2   Find the value of each y when $x = 10$

   a $y = x + 4$ **d** $y = 2x + 3$ **g** $y = 5x - 12$

   b $y = 3x$ **e** $y = 14 - x$ **h** $y = 2(x - 2)$

   c $y = x - 3$ **f** $y = 40 + 3x$

3   Mrs Bujjit is working out the wages at the factory. She uses the formula:
Wages equal hours worked multiplied by rate per hour.

   a   How much does Ellen earn if she is paid £7 an hour and she works for 30 hours?

   b   How much does Francis earn if he works for 25 hours and is paid £8 an hour?

   c   Write out Mrs Bujjit's formula in algebra.

4   Find the value of r when $x = 3$, $y = 4$, $s = 6$ and $t = 0.5$

   a $r = 2x$ **e** $r = \dfrac{s}{3} + 1$ **i** $r = xst$

   b $r = 3y$ **f** $r = \dfrac{2s}{x}$ **j** $r = xy - st$

   c $r = 3x + y$ **g** $r = yt$

   d $r = 2y - t$ **h** $r = 2s - t + y$

5   Find the value of m when $y = -3$

   a $m = 5 + y$ **c** $m = 4y + 6$ **e** $m = 8 - 2y$

   b $m = 3y$ **d** $m = 6 - y$ **f** $m = 5y - 1$

6   In science, Andrea is using the formula $V = IR$
Find V if:

   a $I = 3$ and $R = 42$ **b** $I = 2.5$ and $R = 96$

7   Here are two formulae:

   $y = 2x + 6$ $y = 2(x + 6)$

   If x is 5, work out which formula gives the larger value for y. Show your working.

8   Mary was working through question **2** in her maths lesson.
She said the answer to part **f** was 430.
She explained to her teacher that 40 + 3 = 43, and then 43 × 10 = 430.

What mistake did Mary make?

9   James is doing an experiment in science. He uses the formula

$$\text{Average speed} = \frac{\text{Total distance travelled}}{\text{Total time taken}}$$

He uses this formula to work out that if an object has an average speed of 20 m/s and travels for 10 seconds then it must have travelled 2 m.
What mistake has James made?

10   In geography, Ken is converting degrees Fahrenheit into degrees Celsius.

He uses the formula: $C = \dfrac{5(F - 32)}{9}$

Find C if:

a $F = 68$ **b** $F = 32$

11   In science, Brian is using the formula $S = ut + \frac{1}{2}at^2$

Find s if $u = 3$, $t = 2$ and $a = 4$.

12   Endmoor Cabs use the following formula to work out taxi fares.

> £3.50
> Plus
> £1.20 per mile

a   Jack travels 2 miles. What is Jack's fare?

b   Jane pays £7.10. How many miles has Jane travelled?

c   Jim has £10. What is the maximum distance that he can travel?

Startmoor Cabs have a flat rate of £2 per mile.

d   John needs to travel 10 miles. Which company should he choose?

e   When does it become more expensive to choose Startmoor Cabs?

13   A gym has two schemes that people can choose from, to pay for using the facilities.
Scheme A: Pay £5 per session
Scheme B: Pay an annual fee of £120 and £2 per session.
When would scheme A be better than scheme B?

What if scheme B goes up to £150 and £2 per session while scheme A stays the same?
What if scheme A goes up to £6 per session while scheme B stays the same?

Learn...   11.3 Changing the subject of a formula

Recognising algebraic statements

In this chapter you have used expressions and formulae. You have already solved equations in a previous chapter. You need to be able to identify whether an algebraic statement is an expression, a formula or an **equation**.

Link

See Chapter 5 for more on working with expressions.
See Chapter 7 for a reminder of how to solve equations.

Remember the following about formulae, equations and expressions:

A formula tells you how to work something out. It can be written using words or symbols and will always have an equals sign. There will be at least two letters involved.

For example, here is a formula in words:

Area of a rectangle is equal to length multiplied by width

$A = L \times W$ is the same formula written in symbols, where A stands for area, L for length and W for width. From this formula, you can work out the area of any rectangle if you know its length and width.
You can tell this is a formula, since it tells you what to do with L and W to work out A. There is an equals sign, and there are more than two letters being used.

An equation is two expressions separated by an equals sign. You are often asked to solve an equation, in which case there will be only one letter, but it may appear more than once.

For example, here is an equation in x:

$x + 3 = 7$

In this equation, x is equal to 4. This is the only possible value of x, as any other number added to 3 does not equal 7.
You can tell this is an equation, as there is an expression on each side of the equals sign. There is only one letter involved. It can be solved to find a value of x.

An expression is just a collection of terms. An expression does not have an equals sign.
For example, here is an expression containing x and y terms:

$3x + 2y - 5$

You can tell this is an expression as it is just a collection of terms. There is no equals sign.

Formulae and equations can sometimes look very similar.

Changing the subject of a formula

The **subject** of a formula is the letter on the left-hand side of the equals sign.

P is the subject of the formula $P = 3L + 2$

You can change the subject of this formula to make L the subject.
You will then have a formula telling you what to do to P to work out L.

You use the same strategies that you learned when you solved equations.

> **Hint**
>
> Remember that what you do to one side of an equation, you must also do to the other side.

Example: Make L the subject of the formula $P = 3L + 2$

Solution: $P = 3L + 2$ Write the formula down first.

$P - 2 = 3L + 2 - 2$ Subtract 2 from both sides.

$P - 2 = 3L$

$\dfrac{P - 2}{3} = \dfrac{3L}{3}$ Divide both sides by 3.

$\dfrac{P - 2}{3} = L$

$L = \dfrac{P - 2}{3}$ Rewrite the formula starting with L on the left-hand side.

> **Bump up your grade**
>
> To get a Grade C you need to know how to rearrange formulae.

11.3 Changing the subject of a formula

1 In this chapter you have met these words: **expression**, **formula**, **equation**.
Choose the correct word to describe each of the following.

a $p = a + b + c + d$ **e** $5 = 1 - 2q$

b $2x + 5y - 4z$ **f** $9k - 3$

c $7m + 3n$ **g** $c = 25h - 9$

d $3h = 6$

2 Rearrange the formula $M = n + 42$ to make n the subject.

3 Rearrange each of these formulae to make y the subject.

a $a + y = c$ **c** $e + 2y = f$ **e** $j = 4y - 3k$

b $y - e = d$ **d** $d = h + 3y$ **f** $2m + 6y = n$

4 Rearrange each of these formulae to make x the subject.

a $y = x - 32$ **c** $d = 7x - 60$ **e** $p + sx = t$

b $y = bx$ **d** $kx - p = n$ **f** $y - x = 50$

5 Which of the following is a correct rearrangement of $m = 4x - 3$?

A $x = \dfrac{m - 3}{4}$ **C** $x = \dfrac{m - 4}{3}$ **E** $x = \dfrac{3 - m}{4}$

B $x = \dfrac{m + 3}{4}$ **D** $x = m + \dfrac{3}{4}$ **F** $x = \dfrac{m + 4}{3}$

6 The formula for finding the circumference of a circle is $C = \pi d$

Rearrange this formula to make d the subject.

7 Karen rearranges the formula $y = 4 + x$

She makes x the subject of the formula.

She thinks that the answer is $y + 4 = x$

Is she correct? Give a reason for your answer.

8 Sam rearranges the formula $y = \dfrac{3}{x}$ to make x the subject.

She gets the answer $x = \dfrac{y}{3}$

Is she correct? Give a reason for your answer.

9 Rajesh is using the formula $v = u + at$ in science.

Rearrange the formula to make a the subject of the formula.

10 The formula for finding the volume of a cylinder is

$V = \pi r^2 h$

where V = volume, r = radius and h = height. Write the formula for

finding the height of the cylinder if you know the volume and the radius.

11 Gary needs to change a temperature from °C to °F

He finds this formula: $F = \dfrac{9C}{5} + 32$

Rearrange this formula to make C the subject.

Remember, you must show your working.

12 A holiday is paid for by making a deposit of £300 and then paying instalments of £50 every month.

a How much has been paid after 4 months?

b Write down a formula for the total amount paid, £T, after m months of payments.

c Rearrange your formula to make m the subject.

d Use your rearranged formula to work out the number of months needed to pay for a holiday costing £700. How can you check that your answer is correct?

13 Drive Away car hire base their charges on a fixed daily rate plus an extra allowance for mileage.
Jane wants to hire a car for three days and spend no more than £160.
What is the maximum mileage she can travel?

AQA *Examiner's tip*

Setting up a simple formula can help solve problems like these.

SELF-DRIVE CAR HIRE RATES
—
£50 PER DAY PLUS 5P PER MILE TRAVELLED

11 **Assess** *k!*

G **1** Strickland's corner shop work out their employees' total pay using the formula:

Total pay = Rate per hour × No. of hours + Bonus

Work out the total pay for Harry if he worked for 6 hours at £5 per hour and earned a bonus of £3.

F **2** The fast train from Birmingham to Coventry takes k minutes.
The slow train takes 15 minutes longer.
Write down an expression for the time the slow train takes.

3 Raj got q marks in his maths test.
Sam got three fewer marks than Raj.
Write down an expression for the number of marks Sam got.

4 Find the value of y when $x = 5$

a $y = x + 3$ **b** $y = x - 2$ **c** $y = 3x + 9$ **d** $y = 20 - 3x$

E **5** Write these word formulae using symbols:

a The perimeter of a rectangle is equal to twice the length plus twice the width.

b Kate's weekly wage is equal to £7 for every hour she works.

c Distance travelled is equal to speed multiplied by time.

d The volume of a cuboid is equal to the length multiplied by the width multiplied by the height.

e The length of a rectangle is equal to the area divided by the width.

f Joe delivers pies to customers' homes.
His charge is £5 per pie and he adds on a £2 delivery charge.

6 John uses the formula $P = 2L + 2W$ to find the perimeter of rectangles. The length is L and the width is W. Use John's formula to find the perimeter of a rectangle with

  a length 8 cm and width 5 cm.

  b width 11 m and length 4 m.

7 Mary uses the formula $v = u + at$ in science. Find v if $a = -5$, $u = 24$ and $t = 3$.

8 Jake is having a birthday party.

His Mum buys x packs of balloons and y tweeters.

Each pack of balloons costs 25p and one tweeter costs 14p.

Write down an expression for the total cost of the balloons and tweeters.

9 Which of the words **expression, equation, formula** best describes the following?

  a $m = 4n + p$

  b $3m = 2m + 5$

  c $m + 3n$

10 Karen is using the formula $C = \dfrac{(A + 1)}{9}$

Find C if $A = 26$

11 Make x the subject of the equation $y = 4x - 7$

12 Rearrange the formula $m = 3(C - 2)$ to make C the subject.

AQA Examination-style questions

1 A shopkeeper uses this formula to calculate the total cost when customers pay by monthly instalments.

$$C = d + 24 \times m$$

C is the total cost in pounds.
d is the deposit in pounds.
m is the monthly instalment in pounds.

  a The deposit for a wardrobe is £16.
   The monthly payments are £10.
   What is the total cost? *(2 marks)*

  b How many years does it take to finish paying for goods using this formula? *(1 mark)*

  c The total cost of a sofa is £600.
   The deposit is £120.
   Work out the value of the monthly instalment. *(3 marks)*

  AQA 2009

12 Ratio and proportion

Objectives

Examiners would normally expect students who get these grades to be able to:

D

use ratio notation, including reduction to its simplest form and its various links to fraction notation

divide a quantity in a given ratio

solve simple ratio and proportion problems, such as finding the ratio of one quantity to another

C

solve more complex ratio and proportion problems, such as sharing out a quantity in a given ratio

solve ratio and proportion problems using the unitary method.

Did you know?

It's a goal!

The table shows the number of goals scored by some of Manchester United's players in one season.
It also shows how many games they played in.

Player	Number of goals	Number of games
Michael Carrick	4	28
Michael Owen	8	28
Wayne Rooney	12	30

Who do you think is best at scoring goals?

Key terms

ratio
proportion
unitary method

You should already know:

✓ how to add, subtract, multiply and divide integers

✓ how to simplify fractions.

Learn...   **12.1 Finding and simplifying ratios** (k!)

Here is the table of Manchester United goal scorers again.

Player	Number of goals	Number of games
Michael Carrick	4	28
Michael Owen	8	28
Wayne Rooney	12	30

Ratios can be used to compare numbers and quantities.

There are also other ways such as using fractions or percentages.

A colon is used to separate the parts in a ratio.

For Michael Carrick the ratio of goals to games is $4:28$ $4:28$ is read as '4 to 28'.

You can simplify ratios in the same way as fractions by dividing both parts by the same number.

Michael Carrick's goals to game ratio $= 4:28$

$\div 4$ $\div 4$

$1:7$

This means that on average Michael Carrick scored one goal for every seven games he played.

Michael Owen's goals to game ratio $= 8:28$

$\div 4$ $\div 4$

$2:7$

There are no numbers that divide exactly into 2 and into 7 (apart from 1).

So $2:7$ is the **simplest form** of this ratio.

On average Michael Owen scored two goals for every seven games he played.

Wayne Rooney's goals to game ratio $= 12:30$

$\div 6$ $\div 6$

$2:5$

Or you could divide by 2, then 3.
$12:30 = 6:15 = 2:5$

On average Wayne Rooney scored two goals for every five games he played.

The simplest forms of the ratios are listed in the table.

Player	Simplest form goals : games	1 : n form goals : games
Michael Carrick	$1:7$	$1:7$
Michael Owen	$2:7$	$1:3\frac{1}{2}$
Wayne Rooney	$2:5$	$1:2\frac{1}{2}$

The order of the numbers is very important – so you can tell which are the goals and which are the games.

Dividing the original ratio $12:30$ by 12 gives this.

This is sometimes called a **unitary ratio**.

The simplest forms show that on average Michael Owen scored twice as many goals in the same number of games as Michael Carrick.

It also shows that it took Michael Owen on average seven games to score two goals but Wayne Rooney just five games to score two goals.

Sometimes ratios are divided until one side is 1. This gives the $1:n$ (or the $n:1$) form.
Dividing $2:7$ by 2 gives the ratio $1:3\frac{1}{2}$ (or $1:3.5$). Also $2:5 = 1:2\frac{1}{2}$ (or $1:2.5$).

Example: The total mark for an exam is 80, with 30 marks for Section A and 50 marks for Section B.

Find the ratio of the marks for Section A to the marks for Section B.

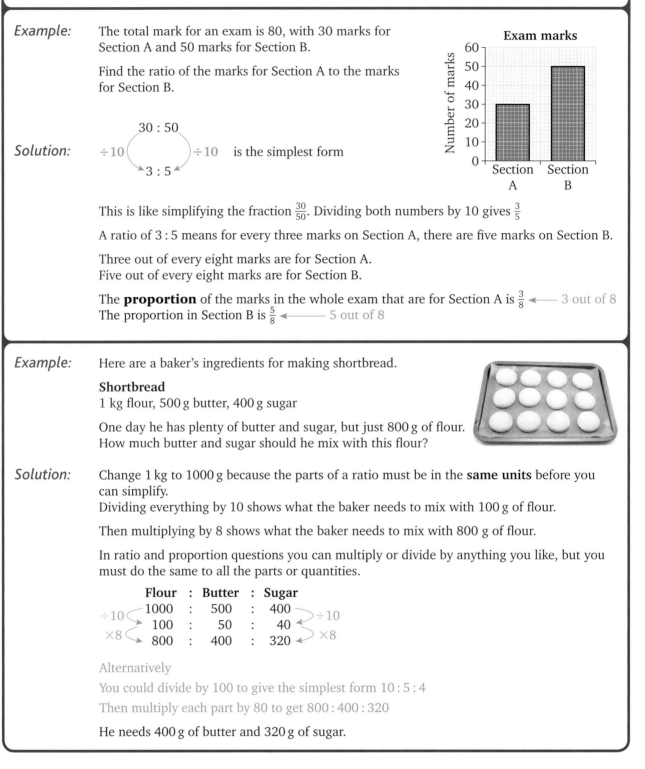

Solution: $\div 10$ 30 : 50 $\div 10$ is the simplest form

       3 : 5

This is like simplifying the fraction $\frac{30}{50}$. Dividing both numbers by 10 gives $\frac{3}{5}$

A ratio of 3 : 5 means for every three marks on Section A, there are five marks on Section B.

Three out of every eight marks are for Section A.
Five out of every eight marks are for Section B.

The **proportion** of the marks in the whole exam that are for Section A is $\frac{3}{8}$ ⟵ 3 out of 8
The proportion in Section B is $\frac{5}{8}$ ⟵ 5 out of 8

Example: Here are a baker's ingredients for making shortbread.

Shortbread
1 kg flour, 500 g butter, 400 g sugar

One day he has plenty of butter and sugar, but just 800 g of flour. How much butter and sugar should he mix with this flour?

Solution: Change 1 kg to 1000 g because the parts of a ratio must be in the **same units** before you can simplify.
Dividing everything by 10 shows what the baker needs to mix with 100 g of flour.

Then multiplying by 8 shows what the baker needs to mix with 800 g of flour.

In ratio and proportion questions you can multiply or divide by anything you like, but you must do the same to all the parts or quantities.

Flour	:	Butter	:	Sugar
1000	:	500	:	400
100	:	50	:	40
800	:	400	:	320

$\div 10$ $\times 8$ $\div 10$ $\times 8$

Alternatively

You could divide by 100 to give the simplest form 10 : 5 : 4

Then multiply each part by 80 to get 800 : 400 : 320

He needs 400 g of butter and 320 g of sugar.

12.1 Finding and simplifying ratios

Practise...

G F E D C

D

1 Write each of these ratios in its simplest form.

   a   4 : 2   **f**   50p : £3.50

   b   4 : 4   **g**   100 cm : 1 m

   c   4 : 6   **h**   350 g : 1.5 kg

   d   4 : 8   **i**   £5 : 45p

   e   4 : 10   **j**   50p : £50

AQA *Examiner's tip*

Write all parts in the same units before simplifying a ratio.

Hint

1 m = 100 cm
1 kg = 1000 g

2 Which of these ratios are equivalent to 1 : 2?

  a £2.50 : £5   **c** 100 : 50   **e** 1 kg : 500 g

  b 50 : 100   **d** 30 min : 1 h   **f** Explain how you can tell that a ratio is equivalent to 1 : 2.

3 Each ratio **a** to **e** is equivalent to 2 : 3. Find the missing number in each ratio.

  a 4 : ☐   **c** ☐ : 6   **e** ☐ : £1.50

  b 12 : ☐   **d** 100 : ☐   **f** Write down three more pairs of numbers or quantities that are in the ratio 2 : 3

4 A film club has 24 females and 18 males.

  a Find the female : male ratio in its simplest form.

  Six more men join the club and ten more women join the club.

  b Is the female : male ratio bigger, smaller, or the same as before?

5 **a** Write down three different pairs of numbers that are in the ratio 1 : 10

  b Explain how you can tell that two numbers are in the ratio 1 : 10

6 The simplest version of all these ratios is 1 : 4. Fill in the gaps.

  a 2 : ☐   **c** ☐ : 28   **e** a : ☐

  b 5 : ☐   **d** ☐ : 3600

7 Find which ratio is the odd one out and explain why.

  a 20 : 25   **c** 8 : 10   **e** 200 : 250

  b $4a : 5a$ **d** 15 : 20

8 School A has 1200 pupils and 80 teachers; School B has 840 pupils and 56 teachers. Do both schools have the same pupil : teacher ratio?

9 The numbers a and b are in the ratio 3 : 4

  a If a is 3, what is b?   **d** If b is 1, what is a?

  b If b is 12, what is a?   **e** If a and b add up to 14, what are a and b?

  c If a is 1, what is b?

10 Here are the ingredients for pastry for a flan for six people.

80 g butter

160 g flour

water to mix

Sophie says, 'So if we make the flan for eight people we will need 100 g of butter and 180 g of flour.'

Is Sophie correct? Explain your answer.

11 On a map, 1 cm represents 10 km.

Which of these is the correct ratio of distance on map : actual distance?

1 : 10; 1 : 100; 1 : 1000; 1 : 10 000; 1 : 100 000; 1 : 1 000 000

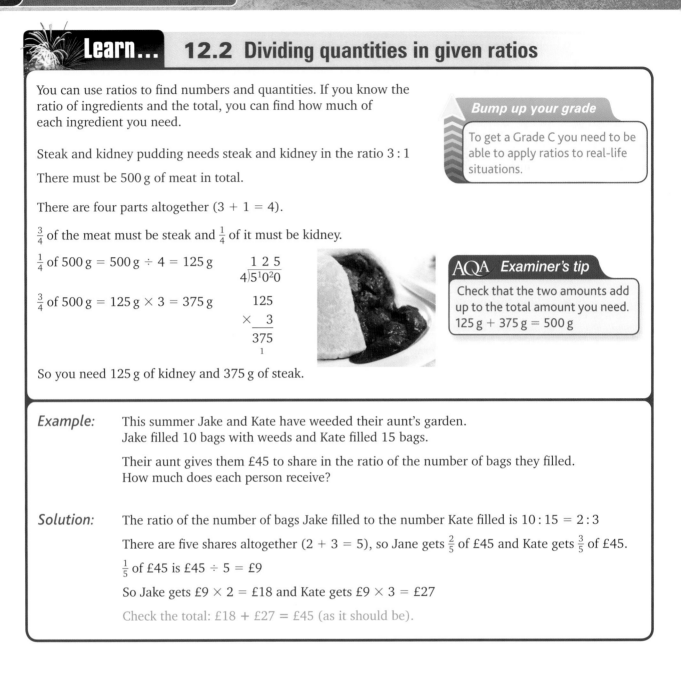

Learn... **12.2 Dividing quantities in given ratios**

You can use ratios to find numbers and quantities. If you know the ratio of ingredients and the total, you can find how much of each ingredient you need.

Steak and kidney pudding needs steak and kidney in the ratio 3 : 1

There must be 500 g of meat in total.

There are four parts altogether (3 + 1 = 4).

$\frac{3}{4}$ of the meat must be steak and $\frac{1}{4}$ of it must be kidney.

$\frac{1}{4}$ of 500 g = 500 g ÷ 4 = 125 g

$$4\overline{)5^{10}0^20}$$

$\frac{3}{4}$ of 500 g = 125 g × 3 = 375 g

$$\begin{array}{r} 125 \\ \times\quad 3 \\ \hline 375 \\ {\scriptstyle 1} \end{array}$$

So you need 125 g of kidney and 375 g of steak.

Bump up your grade

To get a Grade C you need to be able to apply ratios to real-life situations.

AQA *Examiner's tip*

Check that the two amounts add up to the total amount you need.
125 g + 375 g = 500 g

Example: This summer Jake and Kate have weeded their aunt's garden. Jake filled 10 bags with weeds and Kate filled 15 bags.

Their aunt gives them £45 to share in the ratio of the number of bags they filled. How much does each person receive?

Solution: The ratio of the number of bags Jake filled to the number Kate filled is 10 : 15 = 2 : 3

There are five shares altogether (2 + 3 = 5), so Jane gets $\frac{2}{5}$ of £45 and Kate gets $\frac{3}{5}$ of £45.

$\frac{1}{5}$ of £45 is £45 ÷ 5 = £9

So Jake gets £9 × 2 = £18 and Kate gets £9 × 3 = £27

Check the total: £18 + £27 = £45 (as it should be).

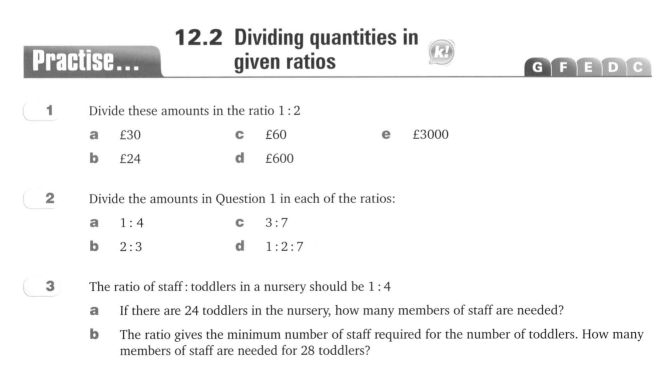

Practise... **12.2 Dividing quantities in given ratios** 🗨️ **G F E D C**

D

1 Divide these amounts in the ratio 1 : 2

  a £30   **c** £60   **e** £3000

  b £24   **d** £600

2 Divide the amounts in Question 1 in each of the ratios:

  a 1 : 4   **c** 3 : 7

  b 2 : 3   **d** 1 : 2 : 7

3 The ratio of staff : toddlers in a nursery should be 1 : 4

  a If there are 24 toddlers in the nursery, how many members of staff are needed?

  b The ratio gives the minimum number of staff required for the number of toddlers. How many members of staff are needed for 28 toddlers?

4   **a**   The ratio of full-colour pages to black and white in a magazine is 3 : 5

      i   If there are 15 full-colour pages, how many black and white pages are there?

      ii   If there are 20 black and white pages, how many pages are there in the magazine altogether?

    b   A different production company has a magazine with 56 full-colour pages and 64 black and white pages. The next edition has 90 pages with the same ratio of full-colour to black and white pages. How many full-colour pages are there?

5   Find the numbers of boys and girls in these schools.

School	Number of students	Boy : girl ratio
School A	844	1 : 1
School B	960	2 : 3
School C	770	3 : 4
School D	810	4 : 5
School E	950	10 : 9

6   **a**   The angles of one triangle are in the ratio 1 : 2 : 3

      i   Find the size of the largest angle.

      ii   What sort of triangle is it?

    b   What sort of triangle is one whose angles are in the ratio 1 : 2 : 6?

> **Hint**
>
> In an **acute-angled** triangle, all three angles are less than 90°.
> A **right-angled triangle** has one angle of 90°.
> An **obtuse-angled triangle** has one angle over 90°.

7   Write an explanation to tell someone how to split a number in the ratio 2 : 3

8   The ratio of fat : sugar : flour in a crumble topping mixture is 1 : 1 : 2

    a   How much flour do you need to make 200 g of crumble topping mixture?

    b   How much crumble topping mixture can you make if you have plenty of flour and sugar but only 30 g of fat?

9   Waseem invested £3500 in a business and Ruksana invested £2500.

    a   At the end of the year, Waseem and Ruksana share the profit of £30 000 in the ratio of their investments. How much does Waseem receive?

    b   Next year, when the profit is shared in the same ratio, Ruksana gets £10 000. What is the total profit?

10   The bar chart shows the pupil : staff ratio in primary schools in various countries. For example, Denmark has the lowest pupil : staff ratio, with approximately 11 students for each teacher.

    a   Approximately how many teachers would there be in a primary school in Denmark with 200 pupils?

    b   Approximately how many teachers would there be in a primary school in Zimbabwe with the same number of pupils?

    c   Approximately how many teachers would there be in a primary school in the United Kingdom with the same number of pupils?

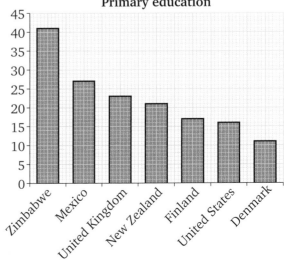

Primary education

Learn... 12.3 The unitary method

The **unitary method** relies on finding a unit amount of one quantity.

If you know that someone can iron six shirts in an hour, you can use the unitary method to work out how long it will take to iron any number of shirts.

Six shirts take one hour, which is 60 minutes.

So one shirt takes 60 minutes ÷ 6 = 10 minutes

You can use the time for one shirt to find the time for any number of shirts by multiplying by the number.

So the time taken for 20 shirts is 20 × 10 minutes = 200 minutes = 3 hours 20 minutes

Example: On a map, 5 cm represents 2 km. Amy measures the distance between two villages on the map as 12 cm. How many kilometres apart are the villages?

Solution: 5 cm represents 2 km

So 1 cm represents $\quad$ 2 km ÷ 5 $\quad = \frac{2}{5}$ km

So 12 cm represents $\quad$ 12 km $\times \frac{2}{5} = \frac{24}{5}$ km

$\qquad\qquad\qquad\qquad\qquad = 4.8$ km

So the villages are 4.8 km apart.

> **Bump up your grade**
>
> To get a Grade C, you should be able to use the unitary method.

Practise... 12.3 The unitary method (k!) $\quad$ G F E D C

1 On a map, 5 cm represents 4 km. What do these represent?

$\quad$ **a** $\quad$ 10 cm $\qquad$ **b** $\quad$ 12 cm $\qquad$ **c** $\quad$ 22 cm

2 Keith irons five shirts in one hour. How long will he take to iron:

$\quad$ **a** $\quad$ 8 shirts $\qquad$ **b** $\quad$ 12 shirts $\qquad$ **c** $\quad$ 2 shirts?

3 Five cups of coffee cost £7.50. How much do three cups of coffee cost?

4 A factory produces 600 bags of crisps in one hour.
How long will it take the factory to produce 1000 bags of crisps?

5 Liz earns £200 for eight hours' work. How much does she earn for 10 hours' work?

6 In a pie chart about people's income, 100 degrees represents 180 people.

$\quad$ **a** $\quad$ How many people does 60 degrees represent? $\qquad$ **b** $\quad$ What angle represents 120 people?

7 70% of a number is 140. What is the number?

8 50 miles is 80 km. 20 miles is how many kilometres?

9 A contract takes six people ten days to complete. How long would four people take to do the same work?

10 x pens cost c pence. How much do y pens at the same price cost?

11 15 hotel housekeepers can clean 20 bedrooms in three hours.
How long will it take ten housekeepers to clean 30 bedrooms?

12 Assess 🗨

D

1 Simplify each of these ratios.

    a $15:12$ **c** $150:250$ **e** $75p:£3$ **g** $250\,g:2\,kg$

    b $25:250$ **d** $£3.50:£7$ **f** $150\,cm:3\,m$

2 Copy and complete this table of pairs of numbers in the ratio $1:3$

1	3
2	
8	
	30
	150
0.5	
n	

3 A school has 55 teachers and 990 pupils.
Express the ratio of teachers to pupils in its simplest form.

C

4 In a hospital A & E unit one month, the ratio of first visits to follow-up visits was $1:24$
There were 4000 visits altogether. How many of these were follow-up visits?

5 In a race, prize money of £150 is split between the runners who came first, second and third in the ratio $3:2:1$. How much does the winner receive?

6 Jane and Bill spend their income on rent, car and other expenses in the ratio $3:1:6$
Their income is £35 000 a year. How much of this do they spend on rent?

7 40% of the pupils in a Year 5 class are girls. What is the girl to boy ratio in this class?
There are 14 girls in the class. How many boys are there?

8 To make sugar syrup, 100 grams of sugar is mixed with 250 ml of water.

    a   How many grams of sugar are mixed with 1000 ml (one litre) of water?

    b   How much water is mixed with 150 grams of sugar?

9 18 carat gold is gold mixed with other metals in the ratio $3:1$.
How much gold is there in an 18 carat gold bracelet weighing 30 g?

10 The table shows the ratio of teachers of different ages in the UK.*

Under 30		30–39		40–49		50 and over
4	:	4	:	7	:	5

    a   What is the ratio
teachers under 30 : teachers aged 30 or over?   *Numbers rounded to nearest integer*

    b   There are approximately 500 000 teachers in the UK. How many of them are under 30?

11 28 pouches of cat food feed two cats for a week. How many pouches do three cats need for four days?

AQA Examination-style questions 🗨

1 A short necklace has 24 gold beads and 16 black beads.
A long necklace has a total of 60 beads.
Both necklaces have the same ratio of gold beads to black beads.
How many black beads are on the long necklace?   *(3 marks)*

                                                       AQA 2004

13 Real-life graphs

Objectives

Examiners would normally expect students who get these grades to be able to:

F

plot points on conversion graphs

read values from conversion graphs

E

read a value from a conversion graph for a negative value

interpret horizontal lines on a distance–time graph

carry out simple interpretation of graphs such as finding a distance from distance–time graphs

D

carry out more advanced interpretation of real-life graphs, such as finding simple average speed from distance–time graphs and recognising when the fastest average speed takes place

construct linear functions from real-life situations and plot their corresponding graphs

C

find the average speed in km/h from a distance–time graph over time in minutes.

- 0.00s

Did you know?

... how important graphs can be?

Graphs that record information such as heart rate, heart beats and blood pressure are very important. Real-life graphs such as these are used in hospitals and can save lives.

You should already know:

✔ how to plot points on a graph

✔ how to draw, scale and label axes

✔ how to complete a table and use it to draw the graph of a straight line

✔ how to plot and interpret a line graph

✔ how to find the gradient of a straight line

✔ how to solve simple problems involving proportion

✔ common units for measuring distance, speed and time.

Learn... 13.1 Conversion graphs and other linear real-life graphs

Conversion graphs are used to convert from one unit of measurement to another.

You need to plot two points to show where to draw your line, and a third point to check.

Example: Shane is converting between euros (€) and pounds (£). He knows that €10 = £8.60

a Draw a conversion graph to convert euros (€) to pounds (£).

b Use the graph to convert
  i €4 to pounds
  ii £7.50 to euros.

Solution

a Choose two points. The easiest are (0, 0) and (10, 8.6). A third point gives a good check. Since €5 = £4.30, (5, 4.3) is an easy point to work out and use. Plot these points and join them with a single straight line.

b i Draw a line from 4 on the euros axis to the line, then from the line to the pounds axis. Read the value from there and write it down. €4 = £3.40

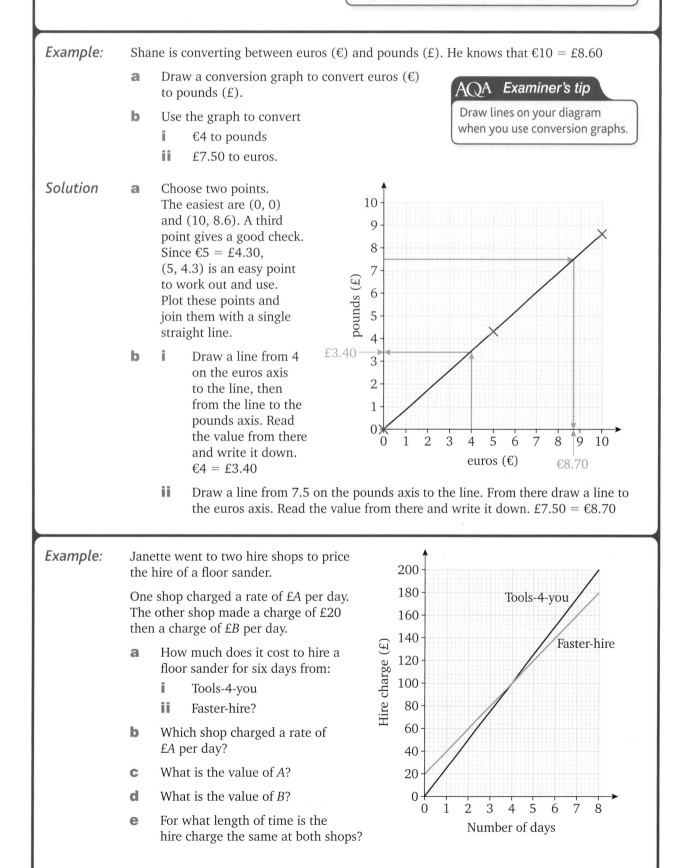

ii Draw a line from 7.5 on the pounds axis to the line. From there draw a line to the euros axis. Read the value from there and write it down. £7.50 = €8.70

Example: Janette went to two hire shops to price the hire of a floor sander.

One shop charged a rate of £A per day. The other shop made a charge of £20 then a charge of £B per day.

a How much does it cost to hire a floor sander for six days from:
  i Tools-4-you
  ii Faster-hire?

b Which shop charged a rate of £A per day?

c What is the value of A?

d What is the value of B?

e For what length of time is the hire charge the same at both shops?

Solution:

a Read up from six days on the horizontal axis to each line, then across from the line to the vertical axis. Write down the values. These are shown on the graph:

    i The blue arrows show the cost of hiring from Tools-4-you is £150.

    ii The red arrows show the cost of hiring from Faster-hire is £140.

b Tools-4-you charge a rate of £A per day. You can tell this because the line for Tools-4-you goes through the origin (0, 0).

c The value of A is £25.
You can read this from the graph – look for the cost of hiring for one day.

d The value of B is £20. You can read the cost of hiring for one day, then subtract the charge of £20. £40 − £20 = £20

e Both shops charge the same for four days.
Look at the graph and find where the lines cross over.
This is where the shops charge the same.

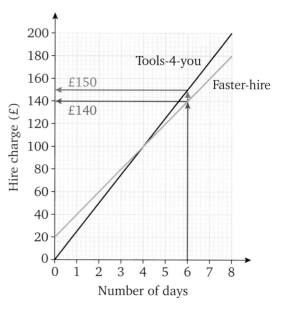

AQA **Examiner's tip**

Draw lines on your graph to show where you are reading from.

Practise... 13.1 **Conversion graphs and other linear real-life graphs** 🔑   G F E D C

F

1 This is a conversion graph for converting between gallons and litres.

  a Use the conversion graph to write the following in litres.

    i 10 gallons   **iii** 12 gallons

    ii 5 gallons   **iv** 2 gallons

  b Use the conversion graph to write the following in gallons.

    i 20 litres   **iii** 45 litres

    ii 70 litres   **iv** 42 litres

  c John bought some petrol at a filling station. He paid £35. Petrol was advertised at £1 per litre. How many gallons of petrol did he buy?

  d Charlie said that 20 gallons were the same as 4.4 litres. He was not correct. What mistake has Charlie made?

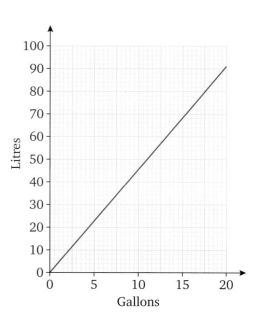

2 **a** Copy the set of **axes** shown on the right, using one large square to represent one unit.

b Rebecca knows that £10 = AUD $19
Use this information to draw a conversion graph for British pounds (£) to Australian dollars (AUD $).

c Use your conversion graph to convert the following amounts into Australian dollars.

i	£4	**iv**	50p
ii	£7	**v**	£9.20
iii	£8.50		

d Use your conversion graph to convert the following into British pounds.

i	A$12	**iv**	A$6.50
ii	A$10	**v**	A$0.50
iii	A$3		

e Rebecca is on holiday in Australia. She sends a postcard home. The charge for postage is A$1.20. She knows that the cost of sending a postcard from home to Australia is 56p. Which is cheaper? Give a reason for your answer.

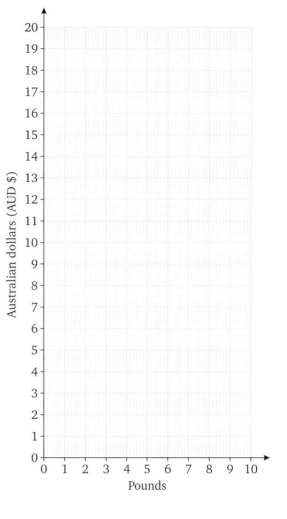

3 **a** At a filling station, diesel costs £1.10 per litre. Copy the table on the right, then use this information to complete it.

Number of litres	0	1	50
Cost		£1.10	

b Copy the set of axes shown below, using one large square to represent ten units.

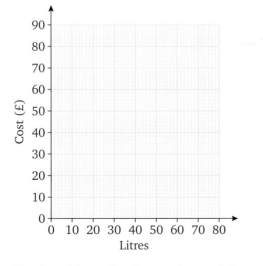

c Use the table to plot three points and draw a line.

d Use your graph to find the cost of:

i	25 litres	**ii**	70 litres	**iii**	65 litres

e Use your graph to find the number of litres bought for:

i	£20	**ii**	£65	**iii**	£16

f Georgina filled up her car with diesel at this filling station. She spent £55.
Use your graph and the graph in Question 1 to find how many gallons of diesel she bought.

F
E

4

a Copy the axes shown on the right.
Use each large square to represent 20 units.

b 0°C is 32°F and 100°C is 212°F. Use this
information to draw a conversion graph for
Celsius (°C) to Fahrenheit (°F).

c Use your conversion graph to convert the
following temperatures to °F.
  i   20°C
  ii 70°C
  iii 90°C
  iv −20°C
  v −60°C

d Use your conversion graph to convert the
following temperatures to °C.
  i   100°F
  ii 80°F
  iii 0°F
  iv −20°F
  v −80°F

e The coldest air temperature ever recorded on
Earth is −89.2°C. It was recorded at Vostok
Station, Antarctica on 21 July 1983. Use your
conversion graph to find this temperature in °F.

f There is one temperature at which the number of °F is the same as the number of °C.
What is this temperature?

g The coldest temperature ever recorded in the UK was −27.2°C in Braemar, Scotland in
January 1982.
Use your conversion graph to find this temperature in °F

h The highest temperature ever recorded in the UK was 38.5°C on 10 August 2003, near
Faversham in Kent. Use your conversion graph to find this temperature in °F

5

a Copy the axes shown on the right.
Use 2 cm to represent ten units.

b Samantha knows that 32 km = 20 miles.
Use this information to draw a conversion
graph for miles to kilometres.

c Use your graph to convert the following
to km.
  i   50 miles
  ii 15 miles
  iii 4 miles

d Use your graph to convert the following to
miles.
  i   20 km
  ii 45 km
  iii 12 km

e Samantha is entering a 10 km race. How far
is this in miles?

f A marathon is 26 miles. Use your graph to
convert this to km.

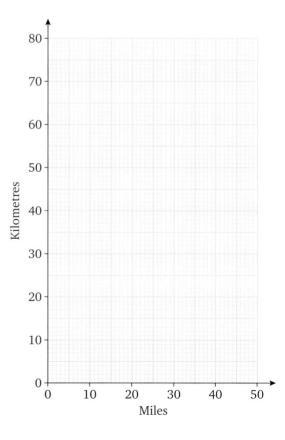

6 Ibrahim is filling his bath with water.

The sketch graph shows how the volume of water in the bath changes.

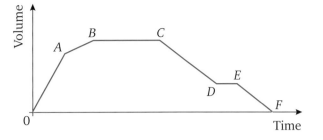

a Which part of the graph shows when the bath was filling up at its fastest? How can you tell?

b What do you think may have happened at A?

c What happened at B?

d Ibrahim cleaned the bath as the bath was emptying. Which part of the graph shows this?
How can you tell?

7 Sharon has a mobile phone.
The only charges she pays are for calls.
The graph shows her monthly charges for calls up to 300 minutes.

a What is her basic monthly charge?

b How many minutes are included in Sharon's basic monthly charge?

c How much does Sharon pay per minute for her other calls?

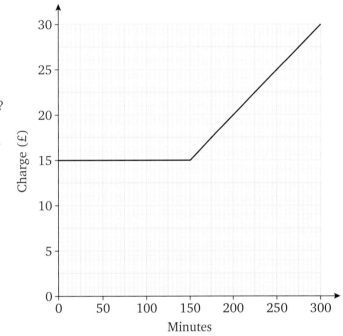

8 Terence the plumber calculates the cost of a job as:
£50 call out fee plus £30 per hour.

a Write down a formula for the price, £P, that Terence would charge for a job taking H hours.

b Make a table for the price he would charge for jobs taking 1, 2, 3, 4 hours.

c Draw a graph showing the price of jobs taking between 0 and 4 hours.

9 Stephen went to two hire shops to hire a bike when he was on holiday.
Fast Bikes charge a £30 hire fee plus £10 per day.
4 Bikers charge a flat rate of £16 per day.

a Write down formulae for the cost, £C, of hiring a bike for n days from:
i Fast Bikes **ii** 4 Bikers.

b How much does it cost to hire a bike for one day from:
i Fast Bikes **ii** 4 Bikers?

c How much does it cost to hire a bike for 2 days from
i Fast bikes **ii** 4 Bikers?

d Draw a graph to show the costs of hiring a bike from each hire shop.

e After how many days is it more expensive to hire from 4 Bikers than from Fast Bikes?

E

D

10 Harriet went to two hire shops to price hiring a mixer.
One shop charged £P per day, the other made a hire charge of £Q plus £R per day.

a How much does it cost to hire a mixer for 5 days from:

i Hire 2 U?

ii Hire Us?

b Which shop charged at a rate of £P per day? How can you tell?

c What are the values of:

i P

ii Q

iii R?

d For what period of time is the hire charge the same from both shops?

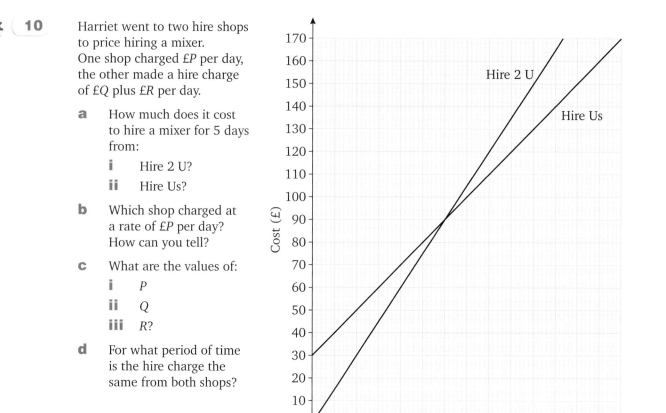

11 A car manufacturer advertises the following for a diesel car:

100 km on only 3.9 litres of fuel, allowing it to cover almost 1500 km without refuelling

Use appropriate conversion graphs to answer the following questions.

a How many gallons does the fuel tank hold when full?

b How many miles per gallon will the car do?

c How much does a full tank of diesel cost for this car?

d How much does each mile cost?

e Jeremy is on holiday from Australia.
How much would Jeremy expect a 50-mile trip in this car to cost in Australian dollars?

Learn... **13.2 Distance–time graphs**

Distance–time graphs tell you about a journey. They are used to compare speeds.

The diagrams show how far Sam and Richard have cycled over a race of 30 metres.

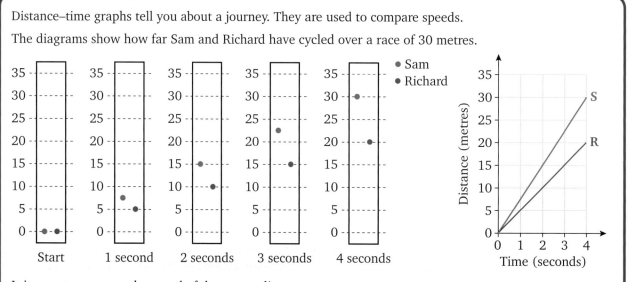

It is easy to compare the speed of the two cyclists.

The vertical **axis** is always distance. The horizontal axis is always time.

The distance is always from a particular point – usually the starting point. The higher up the graph, the further the distance from the starting point.

Time may be the actual time using am and pm or the 24-hour clock, or it could be the number of minutes or hours from the starting point.

If the graph goes back to the horizontal axis, it shows a return to the starting point.

The **gradient** (steepness) of the line is a measure of **speed**. The steeper the line, the faster the speed. A horizontal line represents a speed of zero (i.e. stopped).

In Chapter 10 you learnt to find the gradient of a straight line. The same method is used to find speeds from a distance–time graph, i.e. use:

$$\text{Speed} = \frac{\text{distance travelled}}{\text{time taken}}$$

When the speed is in miles per hour, the distance must be in miles and the time in hours.

To find the average speed for a whole journey, use:

$$\text{Average speed} = \frac{\text{total distance travelled}}{\text{total time taken}}$$

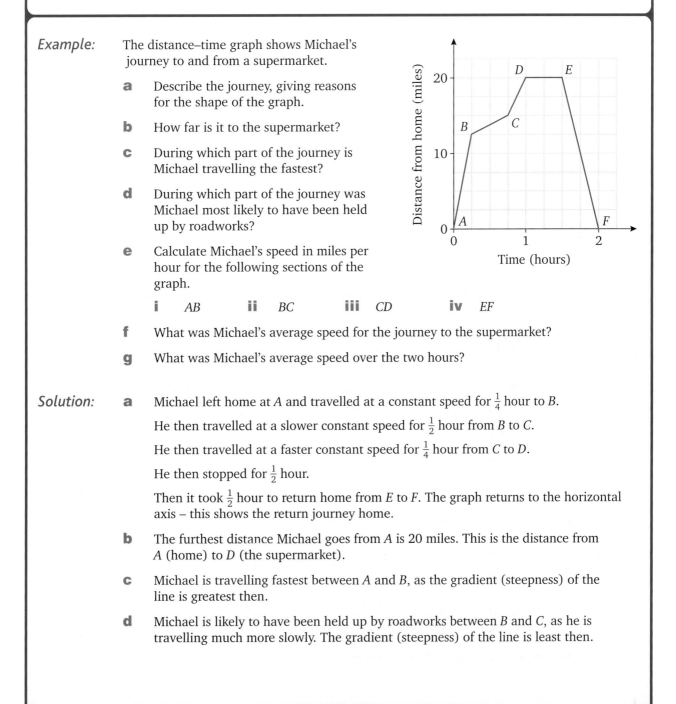

Example: The distance–time graph shows Michael's journey to and from a supermarket.

   a   Describe the journey, giving reasons for the shape of the graph.

   b   How far is it to the supermarket?

   c   During which part of the journey is Michael travelling the fastest?

   d   During which part of the journey was Michael most likely to have been held up by roadworks?

   e   Calculate Michael's speed in miles per hour for the following sections of the graph.

   i   AB   **ii**   BC   **iii**   CD   **iv**   EF

   f   What was Michael's average speed for the journey to the supermarket?

   g   What was Michael's average speed over the two hours?

Solution: **a**   Michael left home at A and travelled at a constant speed for $\frac{1}{4}$ hour to B.

   He then travelled at a slower constant speed for $\frac{1}{2}$ hour from B to C.

   He then travelled at a faster constant speed for $\frac{1}{4}$ hour from C to D.

   He then stopped for $\frac{1}{2}$ hour.

   Then it took $\frac{1}{2}$ hour to return home from E to F. The graph returns to the horizontal axis – this shows the return journey home.

   b   The furthest distance Michael goes from A is 20 miles. This is the distance from A (home) to D (the supermarket).

   c   Michael is travelling fastest between A and B, as the gradient (steepness) of the line is greatest then.

   d   Michael is likely to have been held up by roadworks between B and C, as he is travelling much more slowly. The gradient (steepness) of the line is least then.

e **i** *AB*: In $\frac{1}{4}$ hour, he travels 12.5 miles. In 1 hour he would travel $4 \times 12.5 = 50$ miles
His speed is 50 miles per hour (mph) (as he would travel 50 miles in 1 hour).

   ii *BC*: In $\frac{1}{2}$ hour, he travels 2.5 miles. In 1 hour he would travel $2 \times 2.5 = 5$ miles
His speed is 5 miles per hour (mph).

   iii *CD*: In $\frac{1}{4}$ hour, he travels 5 miles. So in 1 hour he would travel $4 \times 5 = 20$ miles
His speed is 20 miles per hour (mph).

   iv *EF*: In $\frac{1}{2}$ hour, he travels 20 miles. So in 1 hour he would travel $2 \times 20 = 40$ miles
His speed is 40 miles per hour (mph).

f In 1 hour, he travels 20 miles. His average speed is
20 miles per hour (mph).

g For the whole journey the distance travelled is
40 miles (there **and** back). The time taken is 2 hours.

Average speed $= \dfrac{\text{total distance}}{\text{total time}} = \dfrac{40}{2} = 20$ mph

> **AQA** *Examiner's tip*
>
> Remember to divide by the time **in hours** when you find the average speed in miles per hour, or km per hour.

Practise... 13.2 Distance–time graphs G F E D C

1 Sam walks to town to buy a CD and then walks home.
The distance–time graph shows his journey.

a How far does Sam walk altogether?

b Sam stops to talk to a friend on his way into town.
How long does he stop for?

c When is Sam walking at his fastest?
What is his average speed for this part of the journey?

> **AQA** *Examiner's tip*
>
> The horizontal axis has four squares for each hour – this tells you each square is $\frac{1}{4}$ hour. Remember, $\frac{1}{4}$ hour $= 0.25$ hours.

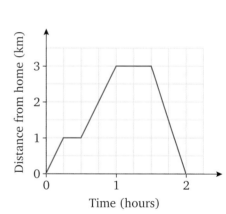

2 Helen completed a short mountain bike trail.
The distance–time graph shows her ride.

a What was the total distance Helen travelled?

b Helen enjoyed the ride, as there was a really fast section.

   i Between which letters on the graph is this indicated?

   ii How long was this section? Give your answer in km.

   iii How long did it take Helen to ride this section?

c There was one very steep uphill section.

   i Between which two letters on the graph is this indicated?

   ii How long was this section?

   iii How long did it take Helen to get up the hill?

d Calculate the average speed in km/h for each of the eight sections of the ride.

e Calculate Helen's average speed for the whole ride.

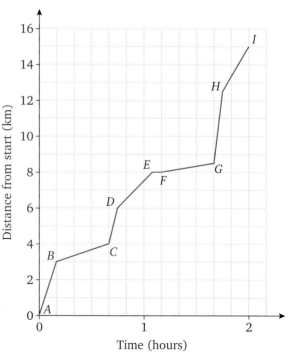

3   A coach travels from Kendal to Birmingham. The journey is shown in the distance–time graph.

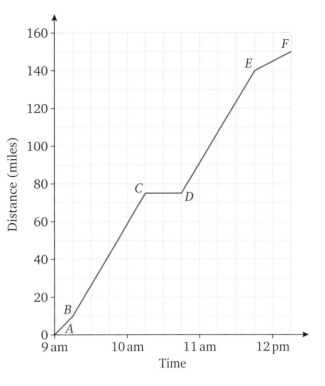

a   The coach stops at some services.

   i   At what time does it stop at the services?

   ii   How long does the coach stop for?

   iii   How far are the services from Kendal?

b   At what stage on the graph does the coach join the motorway?

c   How far is the coach from Birmingham when it leaves the motorway?

d   Work out the average speed in mph for each of the five stages in the journey.

e   Find the average speed of the coach between Kendal and Birmingham.

4   Giovanni goes for a ride on his bike in the country. He starts from the car park and rides for 30 minutes at a steady 12 mph. He then goes up a hill at 8 mph for 15 minutes. At the top he stops to admire the view for 15 minutes. He then rides back down to the car park, which takes him 30 minutes.

Work out Giovanni's average speed in mph for the whole journey.

Hint

You may use a distance–time graph to help.

AQA *Examiner's tip*

When asked to find the average speed in km/h from a distance–time graph with time in minutes, remember to divide by the time taken in hours.

5   The graph below shows the journeys of four students to school in the morning.

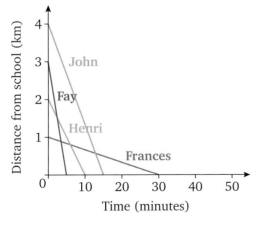

The four students used different ways to get to school.
One student walked, one cycled, one caught the bus and one used the train.

a   How did each student travel to school? Give reasons for your answers.

b   Calculate the speed of each student in km/h.

⚙ **6**   Hamish goes for a ride on his bike.
His journey is shown on the distance–time graph.

a   Hamish fell off his bike at one point in his journey. When do you think this might have been? Explain your answer.

b   When was Hamish going fastest?

c   Describe his journey in words.

d   What was his average speed for the whole journey?

e   What could be changed in the graph for his average speed to have been 5 km/h?

f   How would the graph be different if Hamish had been unable to cycle after he fell off?

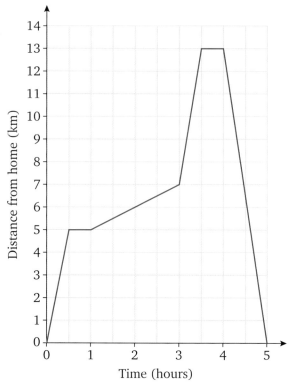

13   Assess 🎮

F   **1**   This is a conversion graph for changing between pints and litres.

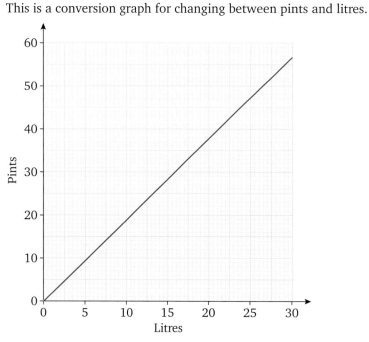

Use the graph to answer the following questions.

a   Convert the following from litres to pints.
        i   10 litres   **ii**   6 litres   **iii**   24 litres   **iv**   18 litres

b   Convert the following from pints to litres.
        i   10 pints   **ii**   45 pints   **iii**   28 pints   **iv**   32 pints

c   Zachary sees two-pint bottles of milk in his local supermarket for 89p. Later he sees one-litre bottles of milk in his corner shop for 86p. Where would you advise Zachary to buy his milk? Give a reason for your answer.

2

a Copy the set of axes below. Use one large square to represent five units.

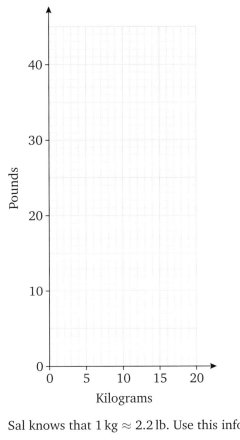

b Sal knows that 1 kg ≈ 2.2 lb. Use this information to draw a conversion graph.

c Use your graph to convert the following weights to kilograms.
    i   10 lb   **ii**   35 lb   **iii** 5 lb   **iv**   42 lb

d Use your graph to convert the following weights to pounds.
    i   8 kg   **ii**   15 kg   **iii** 7.5 kg   **iv**   12 kg

e John estimates his weight to be 140 pounds.
Use your graph to convert his weight to kilograms.

f Sal buys a bag of potatoes weighing 20 lb for £8. The shop advertises a price of £2.10 for 1 kg.
Has Sal been charged correctly? Explain your answer.

3 Colin takes his dog Ben for a walk over Cartmel Fell. The distance–time graph shows his distance from home.

a What time did Colin and Ben set off?

b Colin had his lunch the first time they stopped.
    i   What time did Colin have lunch?
    ii   How long did they stop for lunch?

c What was their average speed in km per hour before lunch?

d On the way back they stopped several times for Colin to admire the view. At what times did Colin make these stops?

e How far did they walk?

f What was their average speed in km/h for the whole walk?

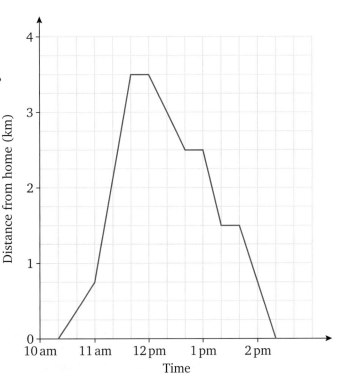

AQA Examination-style questions

1   The diagrams show two journeys *A* and *B*.

   a Which journey has two stops?
Explain your answer.   *(1 mark)*

   b Which part of journey A is the
fastest?

      Copy the diagram and mark
the line with an arrow on the
diagram.   *(1 mark)*

   c The times taken for all three parts of journey B are equal.

      Jill says that the speed for the first and third parts of the journey must be equal.

      Is she correct? Explain your answer.   *(2 marks)*

AQA 2008

2   Motorists should drive with a safe gap between their vehicle and the vehicle in front.
This graph shows the minimum safe gaps between vehicles at different speeds.
Different gaps are recommended for wet roads and dry roads.

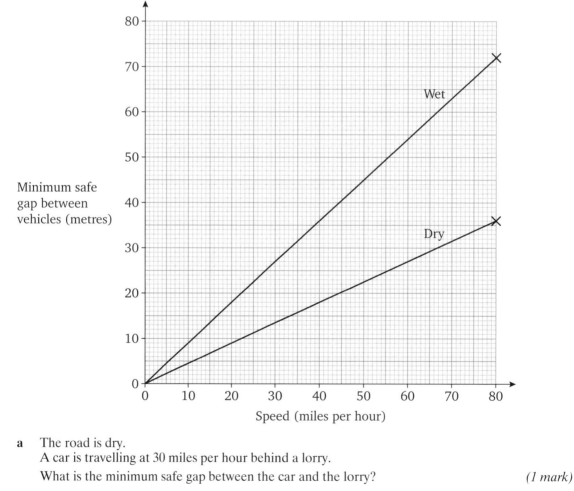

   a The road is dry.
A car is travelling at 30 miles per hour behind a lorry.

      What is the minimum safe gap between the car and the lorry?   *(1 mark)*

   b Tim is driving at 60 miles per hour on a dry road.
He is driving with the minimum safe gap between his car and the car in front.
It starts to rain heavily and both cars slow down to 40 miles per hour.

      Should Tim increase the gap between his car and the car in front to continue driving with the
minimum safe gap?

      You must show clearly how you obtain your answer.   *(3 marks)*

AQA 2008

So far you have covered the following topics:

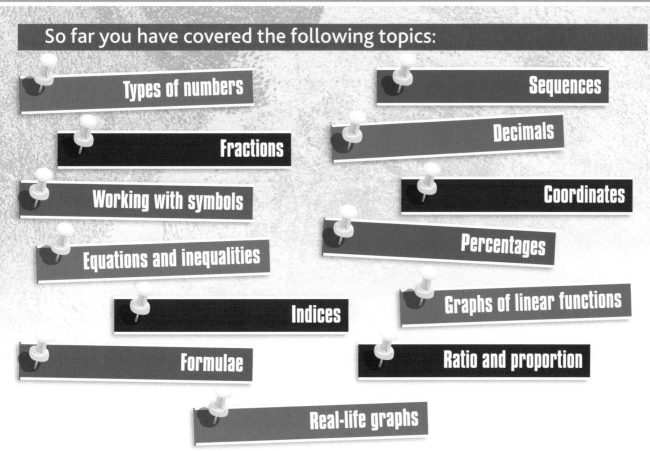

- Types of numbers
- Sequences
- Fractions
- Decimals
- Working with symbols
- Coordinates
- Equations and inequalities
- Percentages
- Indices
- Graphs of linear functions
- Formulae
- Ratio and proportion
- Real-life graphs

All these topics will be tested in this chapter and you will find a mixture of problem solving and functional questions. You won't always be told which bit of maths to use or what type a question is, so you will have to decide on the best method, just like in your exam.

Example: A dressmaker has orders to make three skirts.

One of the skirts is size 8, one size 10 and one size 14.

The dressmaker needs to buy the fabric to make the skirts.

The table shows the length of fabric needed.

Size	8	10	12	14	16	18
Length of fabric	1.9 m	1.9 m	1.9 m	2.2 m	2.2 m	2.2 m

The fabric costs £5.50 per metre.

How much does the dressmaker have to pay for the fabric to make these skirts?

(3 marks)

Solution: Total length of fabric: $1.9 + 1.9 + 2.2 = 6$

So 6 metres of fabric is needed.

Cost of fabric: $6 \times 5.50 = 33$

So the fabric costs £33.00.

> **AQA** *Examiner's tip*
>
> In order to obtain the method marks, it is important to show clearly what you are doing. Do this a step at a time.

> **Mark scheme**
> 1 method mark for $1.9 + 1.9 + 2.2 = 6$
> 1 method mark for 6×5.50
> 1 mark for £33.00.

Example: *ABCDE* is a pentagon.

Find an expression in terms of *x* and *y* for the perimeter of this pentagon.

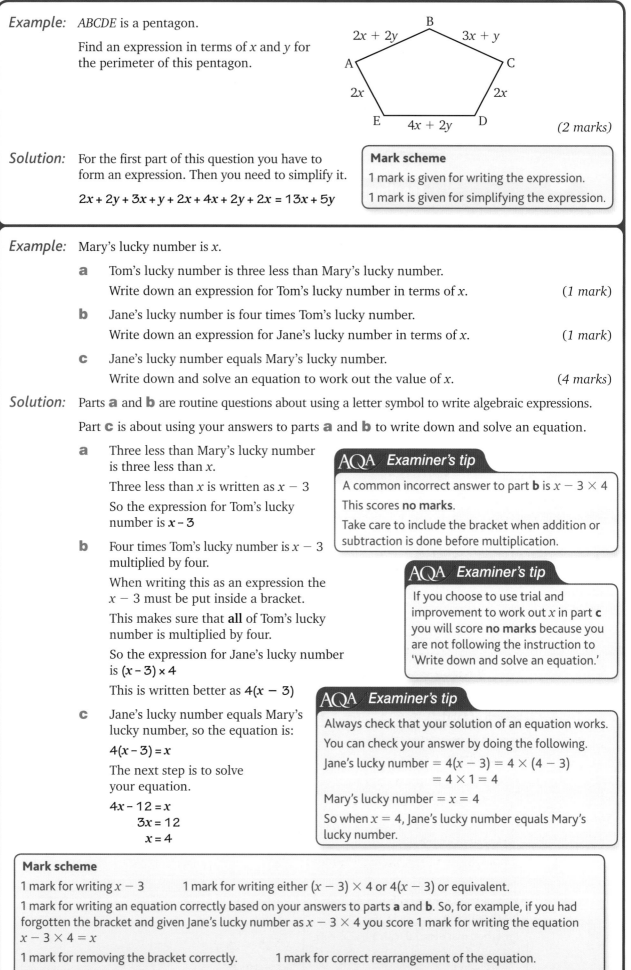

(2 marks)

Solution: For the first part of this question you have to form an expression. Then you need to simplify it.

$2x + 2y + 3x + y + 2x + 4x + 2y + 2x = 13x + 5y$

Mark scheme

1 mark is given for writing the expression.

1 mark is given for simplifying the expression.

Example: Mary's lucky number is *x*.

a Tom's lucky number is three less than Mary's lucky number.

Write down an expression for Tom's lucky number in terms of *x*. *(1 mark)*

b Jane's lucky number is four times Tom's lucky number.

Write down an expression for Jane's lucky number in terms of *x*. *(1 mark)*

c Jane's lucky number equals Mary's lucky number.

Write down and solve an equation to work out the value of *x*. *(4 marks)*

Solution: Parts **a** and **b** are routine questions about using a letter symbol to write algebraic expressions.

Part **c** is about using your answers to parts **a** and **b** to write down and solve an equation.

a Three less than Mary's lucky number is three less than *x*.

Three less than *x* is written as $x - 3$

So the expression for Tom's lucky number is $x - 3$

> **AQA** *Examiner's tip*
>
> A common incorrect answer to part **b** is $x - 3 \times 4$
> This scores **no marks**.
> Take care to include the bracket when addition or subtraction is done before multiplication.

b Four times Tom's lucky number is $x - 3$ multiplied by four.

When writing this as an expression the $x - 3$ must be put inside a bracket.

This makes sure that **all** of Tom's lucky number is multiplied by four.

So the expression for Jane's lucky number is $(x - 3) \times 4$

This is written better as $4(x - 3)$

> **AQA** *Examiner's tip*
>
> If you choose to use trial and improvement to work out *x* in part **c** you will score **no marks** because you are not following the instruction to 'Write down and solve an equation.'

c Jane's lucky number equals Mary's lucky number, so the equation is:

$4(x - 3) = x$

The next step is to solve your equation.

$4x - 12 = x$

$3x = 12$

$x = 4$

> **AQA** *Examiner's tip*
>
> Always check that your solution of an equation works. You can check your answer by doing the following.
> Jane's lucky number $= 4(x - 3) = 4 \times (4 - 3)$
> $\qquad\qquad\qquad = 4 \times 1 = 4$
> Mary's lucky number $= x = 4$
> So when $x = 4$, Jane's lucky number equals Mary's lucky number.

Mark scheme

1 mark for writing $x - 3$　　　　1 mark for writing either $(x - 3) \times 4$ or $4(x - 3)$ or equivalent.

1 mark for writing an equation correctly based on your answers to parts **a** and **b**. So, for example, if you had forgotten the bracket and given Jane's lucky number as $x - 3 \times 4$ you score 1 mark for writing the equation $x - 3 \times 4 = x$

1 mark for removing the bracket correctly.　　　　1 mark for correct rearrangement of the equation.

1 mark for the correct answer.

Consolidation 🔑

1 Here are the heights of five friends.

Alice 1.56 m   Ben 1.53 m   Chloe 1.61 m   Dev 1.62 m   Elly 1.55 m

The friends stand in a queue from tallest to shortest, with the tallest at the back.

a Who is at the back of the queue?

b Who is at the front of the queue?

c Who is in the middle of the queue?

2 Grey squares are used to make patterns.

Pattern 1   Pattern 2   Pattern 3

a Draw Pattern 4.

b How many grey squares are used to make Pattern 5?

c How many grey squares must be added to Pattern 5 if you wanted to make Pattern 8?

d How many grey squares are needed to make Pattern 20?

3 You can make a new number by swapping the positions of two digits in a number.

For example

You are given the number 23 187.

Swap 3 and 8 so that 23 187 → 28 137.

Swap 2 and 7 so that 28 137 → 78 132.

So 23 187 can be changed to 78 132 with two swaps.

a You are given the number 25 147.

    i What is the largest number you can make using **one** swap?

    ii What is the smallest number you can make using **two** swaps?

b You are given the number 3465.

    i The 4 and 3 are swapped.

    Work out the difference between the old number and the new number.

    ii The 5 and the 4 are swapped and then the 4 and the 6 are swapped.

    Work out the difference between the old and the new numbers.

c How many swaps does it take to change the number 1234 to the largest possible **even** number?

Show the numbers you make after each swap.

4 Laura's friend has marked her homework for her but made some mistakes.

Which parts has she marked wrongly?

a $3^2 = 6$ ✓

b $\sqrt{121} = 11$ ✓

c $64 = 4$ ✗

d $\sqrt[3]{125} = 25$ ✗

e $4^3 = 12$ ✗

f $8^2 = 56$ ✓

F **5** Mary goes shopping in a local supermarket.

She buys:
3.5 kg of bananas at £1.20 per kilogram
three pens at £3.75 each
two notebooks at £2.20 each.

She pays with a £20 note.

How much change should she get?

G
F **6** Tim, Stacey and Afzal think of a number and give each other clues about it.
E

a Here are the clues for Tim's number.

CLUE 1 My number is a multiple of 2.
CLUE 2 My number is a multiple of 5.
CLUE 3 My number is a three-digit square number.

Use the clues to work out the number Tim is thinking of.

b Here are the clues for Stacey's number.

CLUE 1 My number is a two-digit factor of 48.
CLUE 2 When I add the digits of my number I do **not** get a factor of 48.

Use the clues to work out the number Stacey is thinking of.

c Here is the clue for Afzal's number.

CLUE 1 My number is a two-digit square number that is also a cube number.

Use the clue to work out the number Afzal is thinking of.

E **7** Amy and Bill take part in a quiz. The rules are:

> For each correct answer score +5 points.
> For each wrong answer score −3 points.

There are five questions in each round.

a After the first round Amy has three questions
correct and Bill has four questions correct.
What are their scores?

b In the second round Bill scores +1. This is added on to his previous score.

Amy wants to have more points than Bill.

What is the least number of questions that Amy must get right in the second
round, so that she has more points?

8 James has 30 marbles.

$\frac{1}{2}$ of the marbles are red.

$\frac{1}{3}$ of the marbles are blue.

The rest of the marbles are green.

Work out the fraction of the marbles that are green.

Give your answer in its simplest form.

9 Shelving units can be bought in the widths shown in the table.

Shelves	Type A	Type B	Type C
Width (metres)	0.65	0.85	1.05
Cost	£37.99	£47.99	£57.99

a The diagram shows two Type A shelves, two Type B shelves and three Type C shelves placed next to each other along a wall.

A	A	B	B	C	C	C

i What is the total length of the shelving units?

ii What is the total cost of the shelving units?

b In a school there is a classroom with one wall of length 6.8 metres.

The school wants to build shelving units along this wall, using as much of the wall as possible.

What shelving units should the school buy to do this?

You **must** show working to justify your answer.

10 Lisa uses 60% of a tin of paint to paint a fence.

What is the smallest number of tins of paint she needs to buy to paint eight fences?

You **must** show your working.

11 The points A, B, C, D and E are plotted on the grid.

a Which two points lie on the line $x = 4$?

b Which two points lie on the line $y = 7$?

c Which two points lie on the line $y = x$?

d Which two points lie on the line $x + y = 11$?

e Which two points lie on the line $y = 2x - 1$?

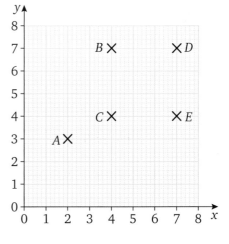

12 **a** Solve the following equations.

i $2x + 11 = 3$

ii $4(y - 2) = 22$

iii $7z + 1 = 10 - 3z$

b In an election, 200 people voted for three candidates.

Sarah wins the election.

She beats Tom by 27 votes and Dilip by 43 votes.

How many votes does Sarah get?

D

13 The Ancient Egyptians used unit fractions like this:

 $= \frac{1}{3}$

A unit fraction has a numerator of 1.

The Ancient Egyptians made other fractions by adding unit fractions together.

For example, $\frac{2}{3}$ can be made by adding $\frac{1}{2}$ and $\frac{1}{6}$

$\frac{1}{2} + \frac{1}{6} = \frac{3}{6} + \frac{1}{6} = \frac{4}{6} = \frac{2}{3}$

> **Hint**
> You can use either any pair of these fractions or all three.

a What fractions can you make by adding $\frac{1}{2}$, $\frac{1}{3}$, and $\frac{1}{4}$?

b An Ancient Egyptian farmer shares five loaves between eight people working in his fields.

   i What fraction of a loaf does each worker get? Give your answer as two unit fractions added together.

   ii Describe how the farmer cuts the loaves to make sure each worker gets an equal share.

c Repeat part **b** for three loaves shared between four workers.

D
C

14 The distances in metres around the outside of a field are shown on the diagram. *BD* is a path across the field.

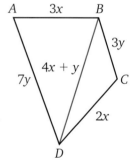

a Adrian walks round the outside of the field from *A* back to *A* again.

Show that the distance that Adrian walks can be written as $5(x + 2y)$ metres.

b Beth walks from *A* to *B*, then along the path *BD* and back along *DA* to *A*.

Write down a simplified expression for how much further Adrian walks than Beth.

c Beth says $y > x$. Is she right? Give a reason for your answer.

15 The nth term of sequence **A** is $2n + 3$

The nth term of sequence **B** is $4n - 1$

a Work out the difference between the 20th term of sequence **A** and the 20th term of sequence **B**.

b Show that 11 is a term in both sequence **A** and sequence **B**.

C

16 A painter needs 15 litres of orange paint.

He decides to make the orange paint by mixing red paint and yellow paint in the ratio 3 : 2

He measures the amount of paint he needs using 200 ml tins.

How many 200 ml tins of each colour of paint are needed?

> **Hint**
> 1 litre = 1000 ml

17 **a** Express 144 as the product of its prime factors. Write your answer in index form.

b Find the highest common factor (HCF) of 30 and 144.

c Find the least common multiple (LCM) of 60 and 144.

18 Dave and Debbie want to go out.

They can go to either the cinema, an exhibition or a show at a theatre.

If they go to these they will need to travel by taxi each way.

They have a choice of two taxi companies.

Use this information to work out which would be the cheapest option for them.

	Distance from home	Ticket price per person
Cinema	8 miles	£8.50
Exhibition	6 miles	£10
Show	5 miles	£15.50

	Charge per trip
A2B Cabs	£2.60 per mile
Sapphire Taxis	£1.20 per mile plus £10

19 Ben sees this entry in an online catalogue.

In addition to the special offer, there is a choice of payment options.

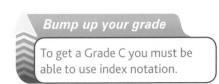

LAPTOP

Ram	3 GB 667 MHz
Hard Drive	320 GB 5400 rpm
Processor	Core 2 Duo 2.26 GHz
Screen	15.4 Inch TFT

£540.00

SPECIAL OFFER
20% OFF MARKED PRICE

		Terms
Option A	Cash payment	10% discount
Option B	6 months to pay	5% deposit Pay the remaining amount monthly in six equal instalments

Ben buys the laptop and decides to pay using **Option B**.

Give full details of what Ben has to pay.

20 Which of these numbers is the smallest?

$a = 4 \times 4^3 \times 4^5$ $b = \dfrac{4^{10}}{4^2}$ $c = \dfrac{4^3 \times 4^6}{4^2}$

You must show all your working.

Bump up your grade

To get a Grade C you must be able to use index notation.

21 A box of sweets falls on the floor.

When the sweets are put together in groups of three, there is one left over.

When the sweets are put together in groups of four, there is one left over.

When the sweets are put together in groups of five, there is one left over.

What is the smallest possible number of sweets that are in the box?

22 **a** Solve the inequality $3x - 5 > 16$

b n is an integer and $-8 \leqslant 2n < 6$

Work out all the possible values for n

AQA Examination-style questions

1   **a**   Giant tubes of fruit gums cost £1.50.
How many giant tubes can Helen buy for £10? *(2 marks)*

   b   Carl spends £2.43 on sweets. He pays with one £10 note.
   i   How much change does Carl receive? *(1 mark)*
   ii   This change is given using the smallest number of notes and coins possible.
How is the change given? *(2 marks)*

AQA 2005

2   **a**   A sequence starts 2, 7, 17, ...
The rule for finding the next term in this sequence is to multiply the previous term
by 2 and then add on 3
Work out the next term. *(1 mark)*

   b   The rule for finding the next term in a different sequence is to multiply the previous
term by 2 and then add on a, where a is an integer.
The first term is 8 and the fourth term is 127

   8   127
Work out the value of a. *(4 marks)*

AQA 2009

3   The graph shows Ben's progress on a sponsored walk.

Distance, km

[Graph with y-axis "Distance, km" marked 0, 5, 10, 15, 20 and x-axis "Time of day" marked 0, 1000, 1100, 1200, 1300, 1400, 1500]

   a   **i**   How long is the walk? *(1 mark)*
   ii   Work out the total time that Ben stops during his walk. *(1 mark)*
   iii   Between which times does Ben walk the fastest? Explain your answer. *(2 marks)*

   b   Sally walks with Ben until they stop at 12:30.
She stops for half an hour longer than Ben but then walks twice as fast as he does.
At what time does Sally catch Ben up?
You **must** show your working. *(3 marks)*

AQA 2004

4   Alan has some unknown weights labelled a and b and some 5 kg and some 10 kg weights.
He finds that the following combinations of weights balance.

   a   Find the value of a.   *(1 mark)*

   b   Find the value of b.   *(2 marks)*

   c   Alan also has some unknown
weights labelled c.

He finds that $5c + 2b = c + 6a$
Work out the value of c.
(4 marks)

AQA, 2007

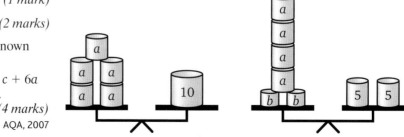

Glossary

< – this inequality sign is used to show that one value is **less than** another. For example, $3 < 8$ means '3 is less than 8'.

≤ – this inequality sign is used to show that one value is **less than or equal to** another. For example, $x \leq 8$ means 'x is any number less than or equal to 8'. In this case x could be any negative number, zero, any positive number below 8 or 8 itself.

> – this inequality sign is used to show that one value is **more than** another. For example, $3 > 1$ means '3 is more than 1'.

≥ – this inequality sign is used to show that one value is **more than or equal to** another. For example, $x \geq -6$ means 'x is any number more than or equal to -6'. In this case x could be any positive number, zero, any negative number above -6 or -6 itself.

amount – the principal + the interest (i.e. the total you will have in the bank or the total you will owe the bank, at the end of the period of time).

ascending – the terms are increasing.

axis (pl. axes) – the lines used to locate a point in the coordinates system; in two dimensions, the x-axis is horizontal and the y-axis is vertical. This system of Cartesian coordinates was devised by the French mathematician and philosopher René Descartes.

balance – how much money you have got in your bank account.

BIDMAS – gives the order in which to carry out operations: brackets, indices, multiplication and division, then addition and subtraction.

brackets – these show that the terms inside should be treated alike, for example $2(3x + 5) = 2 \times 3x + 2 \times 5 = 6x + 10$

coefficient – the number (with its sign) in front of the letter representing the unknown. For example, in $4p - 5$, 4 is the coefficient of p. In $2 - 3p^2$, -3 is the coefficient of p^2.

common factor – factors that two or more numbers have in common.
For example, the factors of 10 are **1**, 2, **5**, 10
the factors of 15 are **1**, 3, **5**, 15
the common factors of 10 and 15 are 1 and 5.

coordinates – a system used to identify a point; an x-coordinate and a y-coordinate give the horizontal and vertical positions.

credit – when you buy goods on 'credit' you do not pay all the cost at once. Instead you make a number of payments at regular intervals, often once a month. NB when your bank account is 'in credit' this means you have some money in it.

cube number – a cube number is the outcome when a number is multiplied by itself and then multiplied by itself again.
125 is a cube number because $5 \times 5 \times 5 = 125$
-5 cubed is $-5 \times -5 \times -5 = -125$

cube root – the cube root of a number such as 125 is the number whose outcome is 125 when multiplied by itself and then multiplied by itself again. The cube root of 125 is 5, as $5 \times 5 \times 5 = 125$

decimal – a number in which a decimal point separates the whole number part from the decimal part, for example 17.46

decimal places – the number of digits to the right of the decimal point. For example, the number 24.567 has three decimal places.

denominator – the bottom number of a fraction, indicating how many fractional parts the unit has been split into. Examples: in the fractions $\frac{4}{7}, \frac{23}{100}, \frac{6}{9}$ the denominators are 7 (indicating that the unit has been split into 7 parts, which are sevenths), 100 and 9.

deposit – an amount of money you pay towards the cost of an item. The rest of the cost is paid later.

depreciation – a reduction in value (for example of used cars).

descending – the terms are decreasing.

difference – the result of subtracting numbers. For example, the difference between 8 and 2 is 6.

digit – any of the numerals from 0 to 9.

directed number – a positive or negative number.

discount – a reduction in the price. Sometimes this is for paying in cash or paying early.

equation – an equation is a statement showing that two expressions are equal, for example $2y - 17 = 15$

equivalent fractions – two or more fractions that have the same value. Equivalent fractions can be made by multiplying or dividing the numerator and denominator of any fraction by the same number. Examples: The fractions $\frac{4}{7}, \frac{8}{14}, \frac{2}{3.5}$ are equivalent and all have the same value.

expand – to remove brackets to create an equivalent expression (expanding is the opposite of factorising).

expression – an expression is a mathematical statement written in symbols, for example $3x + 1$ or $x^2 + 2x$

factor – a whole number which divides exactly into another number with no remainder. For example, the factors of 18 are 1, 2, 3, 6, 9, 18.

factorise – to include brackets by taking common factors (factorising is the opposite of expanding).

formula – a formula shows the relationship between two or more variables. For example, in a rectangle, a formula could be area = length $\times$ width using words, or $A = lw$ using symbols.

gradient – a measure of how steep a line is.
$$\text{Gradient} = \frac{\text{change in vertical distance}}{\text{change in horizontal distance}}$$

highest common factor (HCF) – the highest factor that two or more numbers have in common.
For example, the factors of 12 are **1**, **2**, 3, **4**, 6, 12
the factors of 20 are **1**, **2**, **4**, 5, 10, 20
the common factors are 1, 2, 4
the highest common factor is 4.

horizontal axis – the *x*-axis goes across the page – the *horizontal* axis.

improper fraction – a fraction with a numerator greater than its denominator.

index – the index (or power or exponent) tells you how many times the base number is to be multiplied by itself.

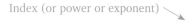

Index (or power or exponent)

$$5^3$$

Base

5^3 tells you that 5 (the base number) is to be multiplied by itself 3 times (the index or power or exponent). So $5^3 = 5 \times 5 \times 5$

indices – the plural of index – see **index**.

inequality – statements such as a $c > b$, $a < b$ or $a \leqslant b$ are inequalities.

integer – any positive or negative whole number or zero, for example $-2, -1, 0, 1, 2, \ldots$

interest – the money paid to you by a bank or building society when you save your money with them. NB it is also the money you pay for borrowing money from a bank.

inverse operation – the opposite operation. For example, subtraction is the inverse of addition. Division is the inverse of multiplication.

least common multiple (LCM) – the least (or lowest) multiple that two or more numbers have in common. For example, the multiples of 4 are 4, 8, **12**, 16, 20, **24**, 28, 32, **36**, …
the multiples of 6 are 6, **12**, 18, **24**, 30, **36**, …
the common multiples are 12, 24, 36
the least common multiple is 12.

like terms – $2x$ and $5x$ are like terms. xy and yx are like terms.

linear – describes an equation, expression, graph, etc. where the highest power of a variable is 1; for example, $3x + 2 = 7$ is a linear equation but $3x^2 + 2 = 7$ is not.

linear sequence – in a linear sequence, the differences are all the same.

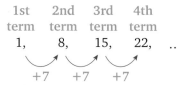

1st term	2nd term	3rd term	4th term	
1,	8,	15,	22,	…

+7 +7 +7

This is a linear sequence.

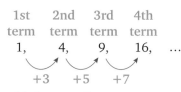

1st term	2nd term	3rd term	4th term	
1,	4,	9,	16,	…

+3 +5 +7

This is not a linear sequence.

line segment – a portion of a straight line, defined by the end points.

midpoint – the point in the middle of a given line, or line segment.

mixed number – a fraction that has both a whole number and a fraction part. Examples: $1\frac{4}{7}, 3\frac{1}{2}, 5\frac{3}{4}$

multiple – the multiples of a number are the products in its multiplication table. For example, the multiples of 7 are 7, 14, 21, 28, 35 …

negative number – a number less than zero. For example $-2, -18\frac{3}{4}, -35.8, -256$.

'nth' term – this phrase is often used to describe a 'general' term in a sequence. The *n*th term is sometimes called the position-to-term rule.

numerator – the top number of a fraction, indicating how many parts there are in the fraction. Examples: in the fractions $\frac{4}{7}, \frac{23}{100}, \frac{6}{9}$ the numerators are 4, 23 and 6.

operation – a rule for combining two numbers or variables, such as add, subtract, multiply or divide.

origin – the point $(0, 0)$ on a coordinate grid.

percentage – 'the number of parts per hundred'. For example, 15% means $\frac{15}{100}$

place value – the value that a digit has depending on its position in a number. For example, in the number 25 674 the value of the digit 7 is 70 and the value of the digit 5 is 5000.

positive number – a number greater than zero. For example $+2, +18\frac{3}{4}, +35.8, +256$.

power – see **index**.

prime number – a natural number with exactly two factors.
The first seven prime numbers are:

2	3	5	7	11	13	17
Factors	Factors	Factors	Factors	Factors	Factors	Factors
1 & 2	1 & 3	1 & 5	1 & 7	1 & 11	1 & 13	1 & 17

1 is not a prime number because it has only one factor. 2 is the only even prime number.

principal – the initial amount of money put into the bank (or borrowed from the bank).

product – the result of multiplying numbers. For example, the product of 8 and 2 is 16.

proportion – if a class has 10 boys and 15 girls, the proportion of boys in the class is $\frac{10}{25}$ (which simplifies to $\frac{2}{5}$). The proportion of girls in the class is $\frac{15}{25}$ (which simplifies to $\frac{3}{5}$). A ratio compares one part with another; a proportion compares one part with the whole.

quadrant – the axes divide the page into four *quadrants*.

quotient – the result of dividing numbers. For example, when 8 is divided by 2, the quotient is 4.

rate – the percentage at which interest is added.

ratio – a ratio is a means of comparing numbers or quantities. It shows how much bigger one number or quantity is than another. If two numbers or quantities are in the ratio $1:2$, the second is always twice as big as the first. If two numbers or quantities are in the ratio $2:5$, for every two parts of the first there are five parts of the second.

reciprocal – any number multiplied by its reciprocal equals 1. 1 divided by a number will give its reciprocal; for example, the reciprocal of 3 is $\frac{1}{3}$ because $3 \times \frac{1}{3} = 1$.

recurring decimal – a decimal number in which a number or group of numbers keeps repeating. You write a recurring decimal like this: $0.\dot{3}$ (the dot shows which number is recurring). If it is a group of numbers that repeats, you put the dot over the first and last digits.

round – give an approximate value of a number; numbers can be rounded to the nearest 1000, nearest 100, nearest 10, nearest integer, significant figures, decimal places, etc.

sequence – a sequence is a list of numbers or diagrams which are connected in some way.

significant figures – the digits in a number. The closer a digit is to the beginning of a number, the more important or significant it is; for example, in the number 23.657, 2 is the most significant figure.

simplify – to make simpler by collecting like terms.

solve/solution – the value of the unknown quantity. For example, if the equation is $3y = 6$, the solution is $y = 2$

speed – speed is the gradient of a line on a distance–time graph. It is found using the same method as in Chapter 10. For distance–time graphs this is:

$$\text{Speed} = \frac{\text{distance travelled}}{\text{time taken}}$$

square number – a square number is the outcome when a number is multiplied by itself.
16 is a square number because $4 \times 4 = 16$
-4 squared is $-4 \times -4 = 16$ also.

square root – the square root of a number such as 16 is the number whose outcome is 16 when multiplied by itself. The square root of 16 is 4, as $4 \times 4 = 16$.
Also, the square root of 16 is -4, as $-4 \times -4 = 16$

subject – the subject of the formula $P = 2(l + w)$ is P because the formula starts '$P = ...$'

substitute – in order to use a formula to work out the value of one of the variables, you replace the letters in the formula with numbers. This is called substitution.

substitution – putting number values into an expression or formula.

sum – the result of adding numbers. For example, the sum of 8 and 2 is 10.

symbol – algebra uses symbols to form expressions and make statements, and letters are often used to represent numbers.

term – a number, a variable, or the product of a number and one or more variables, e.g. 5, x, $3x$, $2xy$, x^2

term-to-term – the rule which tells you how to move from one term to another.

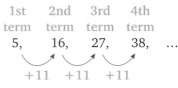

The rule to find the next number in the sequence is $+11$. The rule is called the term-to-term rule.

unitary method – a way of calculating quantities that are in proportion. For example, if six items cost £30 and you want to know the cost of ten items, you can first find the cost of one item by dividing by 6, then find the cost of ten by multiplying by 10.

unitary ratio – a ratio expressed in the form $1:n$ or $n:1$. This form enables comparisons to be made such as school staff : student ratios expressed as $1:12.3$ or $1:13.5$.

unknown – the letter in an equation that represents a quantity which is 'unknown'.

unlike terms – $2x$ and $5y$ are unlike terms. x and x^2 are unlike terms.

value – letters in a formula represent values. The given value of a letter is substituted into a formula to form an equation.

variable – a symbol representing a quantity that can take different values, such as x, y or z.

VAT (Value Added Tax) – this tax is added on to the price of some goods or services.

vertical axis – the y-axis goes up the page – the *vertical* axis.

Index